AQUATIC MITES: FROM GENES TO COMMUNITIES

AQUATIC MITES FROM GENES TO COMMUNITIES

Editor
HEATHER C. PROCTOR
University of Alberta, Edmonton, Canada

Reprinted from *Experimental and Applied Acarology*
Volume 34 Numbers 1–2, 2004

Kluwer Academic Publishers
Dordrecht / Boston / London

A C.I.P. Catalogue record for this book is available from the Library of Congress.

ISBN 1-4020-2703-6

Published by Kluwer Academic Publishers,
P.O. Box 17, 3300 AA Dordrecht, The Netherlands.

Sold and distributed in the U.S.A. and Canada
by Kluwer Academic Publishers,
101 Philip Drive, Norwell, MA 02061, U.S.A.

In all other countries, sold and distributed
by Kluwer Academic Publishers,
P.O. Box 322, 3300 AH Dordrecht, the Netherlands.

Printed on acid-free paper

Printed in the Netherlands

Cover photograph: With their often brilliant colours, water mites are strikingly conspicuous compared to most other freshwater invertebrates. Many species, such as *Limnochares americana* Lundblad (Limnocharidae), are bright red, and are also distasteful to fish. The origin and function of this apparent aposematism are discussed in Proctor & Garga in this volume. Photo by Heather Proctor; *L. americana* specimens from Hasse Lake, Alberta, Canada.

TABLE OF CONTENTS

Experimental and Applied Acarology **34:** 1–2, 2004.

Aquatic mites: from genes to communities – an introduction

Although chelicerates originated in the sea, the vast majority of extant species are terrestrial. Since the invasion of land, only one species of spider has taken up the old subaquatic life (*Argyroneta aquatica*), and there are a few intertidal pseudoscorpions that can withstand inundation. It is among the mites that one finds the greatest number of species, indeed, of entire superfamilies, that have taken up a watery existence. Each aquatic taxon has close terrestrial relatives, and sometimes displays rudiments of terrestrial adaptations (e.g. closed stigmatal openings), and so it seems that invasion of the aquatic environment by mites has occurred repeatedly. Approximately 7000 species from the Mesostigmata, Astigmata, Oribatida and especially the Prostigmata, now live in marine and freshwater habitats. There are even a few ticks that parasitize marine iguanas and sea snakes! In part because of their unusual lifestyle, and likely also because few have any obvious agricultural or medical importance, aquatic mites are usually relegated to habitat-specific journals together with various other aquatic invertebrates, and have rarely appeared in the pages of *Experimental and Applied Acarology*.

We hope that this special issue will help to bring aquatic mites into 'mainstream' Acarology. It is the product of a brief conversation between HP and *EAA*-editor Jan Bruin at the poster session of the International Congress of Acarology in Mérida, Mexico, in September 2002. Jan asked Heather whether she would be interested in acting as guest editor for a special issue of *Experimental and Applied Acarology* dealing with aquatic mites. Saying 'yes' required but a moment's consideration; bringing the issue into fruition has been a rather longer process! But the result was worth it. Here are 12 new papers that cover a wide range of taxa and a great diversity of themes. Marine and freshwater oribatids are featured in papers by Marshall and Convey and by Søvik, where we learn that the Arctic species *Ameronothrus lineatus* takes 5 years to reach maturity (twice as long as the average lifespan of many arctic rodents!). A summary of current knowledge about the distribution and ecology of marine halacarids is presented by Bartsch; this information will be extremely valuable given the increasing transportation of marine taxa by shipping, and the importance of assigning point of origin to animals found in ballast water. The remaining papers deal with the most species-rich group of aquatic mites, the 'true' water mites (Prostigmata: Parasitengona). These range from studies of genes (Rey et al., Edwards et al.), larval parasitism (Forbes et al., Martin), mating behaviour (Smith and Florentino), evolution of warning colouration (Proctor and Garga), and relationship between environmental factors and community composition (Di Sabatino et al., Boulton et al., Goldschmidt). These articles raise as many interesting questions as they answer, and should provoke more studies of the biology of freshwater and marine Acari.

I thank the authors and the many excellent reviewers who helped to create this special issue.

Heather Proctor
Department of Biological Sciences
University of Alberta
Edmonton
Alberta T6G 2E9
Canada
e-mail: hproctor@ualberta.ca

Experimental and Applied Acarology **34:** 3–20, 2004.

The biology and life history of arctic populations of the littoral mite *Ameronothrus lineatus* (Acari, Oribatida)

GULDBORG SØVIK
Institute of Marine Research, Tromsø branch, P.O. Box 6404, N-9294, Tromsø, Norway; (e-mail: guldborg.soevik@imr.no; phone: +47-77-60-9753; fax: +47-77-60-9701)

Key words: Latitudinal distribution, Life cycle, Microevolutionary adaptations, Multi-instar aggregations, Population dynamics, Reproductive biology, Temperature

Abstract. The present study attempts to elucidate possible microevolutionary adaptations of life-history traits of high-latitude populations of the holarctic, littoral oribatid mite *Ameronothrus lineatus* by comparing arctic and temperate populations. Additionally, the paper provides an overview of the limited research on general ecology and population biology of arctic populations. In the Arctic the larviparous *A. lineatus* has a 5-year life cycle (larva-to-larva), and adults survive a further 2–3 years. High survival to maturity is consistent with a low lifetime reproductive output of ca. 20 larvae. The life history can be regarded as an extreme version of the typical oribatid life history. However, several life-history features suggest specific adaptations of arctic populations. In particular, the pre-moult resting stage is synchronized with the warmest part of the arctic summer, which shortens this vulnerable part of development. High reproductive investment by females at relatively low temperatures may represent a physiological adaptation to the cool arctic summer. Finally, prolonged cold exposure positively affects reproduction and survival the following summer, suggesting adaptation of the species to the highly seasonal arctic environment. On the other hand, the ability of all life-cycle stages to overwinter, and a flexible life history with the species being able to take advantage of favourable climatic conditions to accelerate development and larviposition, seem to be ancestral features. Thus, the success of *A. lineatus* in arctic habitats is probably attributable to a combination of derived and ancestral life-history traits. Studies of conspecific temperate populations are required to elucidate further local adaptations of arctic populations.

Introduction

The diversity of the arctic arthropod fauna is fairly well known (Danks 1981; Coulson and Refseth 2004), but information on population biology and life-history strategies of component species is fragmentary (Danks 1981, 1999). Central to our understanding of terrestrial polar arthropods is knowledge about possible life-history adaptations to the harsh polar environment. Many of the typical life-history traits of polar arthropods (e.g., extended life span and reduced reproductive output) agree with predictions from adversity selected life-history strategies (Convey 1996). However, many of these traits must be viewed as plesiotypic (ancestral) characters, enabling successful colonization of the arctic environment, or consequences of ecological or physiological limitations (Danks 1981; Norton 1994; Convey 1996).

Oribatid mites are usually among the most abundant and species-rich arthropods in arctic soils (Behan 1978; Danks 1981), however, little information exists about their demography and life histories in cold regions. In the Antarctic only one species, *Alaskozetes antarcticus* (Michael), has been studied in detail (Block and Convey 1995), whereas the holarctic *Ameronothrus lineatus* (Thörell 1871) is the only oribatid mite inhabiting arctic regions, for which detailed knowledge on life history and demography exists (Søvik et al. 2003; Søvik and Leinaas 2003a, b). The latter species is found in littoral habitats across a wide latitudinal gradient extending from warm-temperate to high-arctic regions. This makes it a good model organism for studies of local adaptation, as population-specific differences in life history can be examined across a correspondingly wide environmental gradient. Warm-temperate coastal regions have a uniform climate with precipitation distributed evenly throughout the year. Growing seasons are long whereas winters are short and mild (average temperatures of 0–3 °C in December–February) (www.metoffice.gov.uk; www.dwd.de). In comparison, summers in the high Arctic are short (2–4 months) and cold with restricted time for reproduction and completion of life cycles. The winter is long with low soil temperatures (0 to −30 °C) (Coulson et al. 1995, 2000), depending on air temperature and snow cover. Arctic habitats are generally arid because of low precipitation and high winds (Danks 1981, 1999).

The present study attempts to elucidate possible local life-history adaptations of *A. lineatus* to the arctic environment by comparing arctic and temperate populations. Additionally, the paper provides an overview of published research on the general biology of arctic populations of *A. lineatus*.

Taxonomy and distribution

Ameronothrus lineatus (Figure 1), first described from Svalbard as *Eremaeus lineatus* (type species), belongs to Ameronothridae Willmann 1931, which comprises species found in intertidal and terrestrial habitats in both hemispheres (Schulte and Weigmann 1977). *Ameronothrus* Berlese 1896 is predominately northern. The 10 species (Gilyarov 1975; Schubart 1975; Weigmann and Schulte 1975) are mainly aquatic or semiaquatic, being distributed on arctic, temperate and subtropical coastlines (Schuster 1966, 1988; Schulte et al. 1975; Schulte and Weigmann 1977).

The latitudinal distribution of *A. lineatus* stretches from southern England to the high Arctic, covering almost 30° of latitude (Colloff 1984). In North America the equally wide distribution is shifted southwards into northern California (Schulte et al. 1975; Schulte 1978). No reports exist from the Canadian high Arctic (Danks 1981), probably due to limited sampling in those inaccessible regions. Thus, *A. lineatus* has so far been found in Iceland, the Faroe Islands, Greenland, Jan Mayen, Svalbard, Novaja Semlja, arctic Siberia,

Figure 1. Ameronothrus lineatus. Lateral view of a female (from Colesbukta, Spitsbergen).

Fennoscandia, Great Britain, Ireland, Germany, Canada, Alaska, and California (Hammer 1944; Strenzke 1955; Weigmann and Schulte 1975; Behan 1978; Danks 1981; Karppinen and Krivolutsky 1982; Colloff 1984; Coulson and Refseth 2004). It is the only ameronothrid reported from the high-arctic Svalbard (Coulson and Refseth 2004), where it has been sampled on Bear Island and along the west and north coast of Spitsbergen (Summerhayes and Elton 1923, 1928; Karppinen 1967; Søvik et al. 2003). The life-history studies referred to in the present work were carried out using specimens from Adventdalen (78°0′N 15°30′E) and Colesbukta (78°5′N 14°57′E) on West Spitsbergen.

Life history

Life-cycle stages

Oribatid mites have six life-cycle stages: pre-larva, larva, proto-, deuto-, tritonymph and adult. The pre-larva does not hatch from the egg (Norton 1994; Walter and Proctor 1999), and in *Ameronothrus* is seen as a layer of cuticle within the eggshell (Schubart 1975). The majority of oribatid species deposit eggs, but some retain the progeny until the larval stage (larviparity) (Norton 1994). This seems to be a plesiotypic trait of the genus *Ameronothrus* (Schubart 1970, 1975; Weigmann and Schulte 1975; Pugh and King 1986; Tilrem 1994; Bücking et al. 1998). Strictly speaking, *A. lineatus* is ovoviviparous as the larvae hatch from the eggs some hours after deposition (Haq et al. 1991; Søvik

2003). The hexapod larvae are easy to separate from the other octopod stages, whereas the nymphal stages are distinguished based on the number of genital papillae (1, 2, 3 for proto-, deuto-, and tritonymphs respectively) or genital setae (1, 3, 5, respectively (Weigmann and Schulte 1977)). Living juveniles are staged based on size (Søvik and Leinaas 2003a). Before moulting, juveniles enter a quiescent stage during which they become turgid and completely immobile for several days to weeks (G. Søvik, unpubl. data). Juveniles and adults are distinguished by the surface of the notogaster, with longitudinal ridges in adults and pleats in juveniles (Schubart 1975; Weigmann and Schulte 1977). The spermatopositor is unusually long compared with most oribatid species (Schubart 1975) and, together with the long ovipositor, makes sexing adults easy. Live adults are sexed by the gender-specific size difference and the longer legs relative to body size of the males (Schubart 1975). *A. lineatus* is a large species (adult length ca. 0.75–0.90 mm), which makes it relatively easy to handle and culture.

Life cycle

Initiation of the first moult after winter is highly synchronous with 80–90% of the juvenile instars in the pre-moult quiescent stage in early summer (Figure 2) (Søvik et al. 2003; Søvik and Leinaas 2003a). Søvik et al. (2003) suggested that the synchronization is a phenological strategy to time quiescence and moulting with the warmest part of the arctic summer, hence shortening this vulnerable, inactive stage, which may last a quarter to a third of development (Luxton 1981a). The long winter probably acts as a synchronizing agent, with phero-

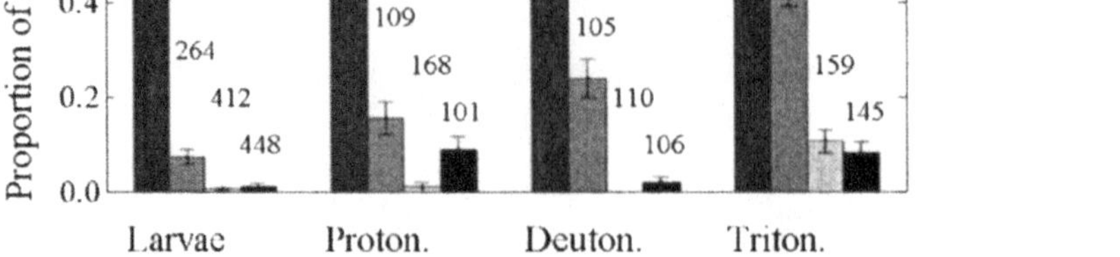

Figure 2. Moult synchronization in *A. lineatus* in field microcosms in Adventdalen, West Spitsbergen (after Søvik et al. 2003). Data are proportions (±SE) of the juvenile instars found in the pre-moult, quiescent stage at different sampling dates summer 1998. Proportions on June 23 were equal to zero. Sample sizes are given in the figure.

mones possibly increasing the synchronization, as was indicated by a higher degree of moult synchronization within than between culturing boxes (Søvik and Leinaas 2003a). A synchronous moult in early summer was also observed in subalpine populations of the peatland species *Trhypochthoniellus setosus* Willmann (Kuriki 1995) and the Antarctic *Alaskozetes antarcticus* (Convey 1994a).

In average arctic summers the majority of juveniles moult once (Søvik and Leinaas 2003a). As females start larvipositing the second adult summer (Søvik and Leinaas 2003b), this gives a 5-year life cycle larva-to-larva. Oribatid life cycles of similar lengths are known from the Antarctic (5 years, Convey 1994a; Marshall and Convey 1999), and alpine regions (3–5 years, Schatz 1985; >4 years, Soma 1990), whereas Mitchell (1977) suggested a 3-year life cycle for *Ceratozetes kananaskis* Mitchell in a boreal forest. In comparison, temperate populations of *A. lineatus* have a generation time of only 1 year (Bücking et al. 1998). Most temperate oribatid species also have life cycles lasting 1–2 years (Luxton 1981a, b; Norton 1994). The prolonged life cycle on Svalbard probably results from short growing seasons and low temperature budgets (Block and Convey 1995), but may also be explained by the suggested synchronized timing of moulting to the warmest part of the summer, with juveniles normally not being able to moult more than once per summer due to climatic constraints (Søvik et al. 2003). Thus, each active stage in the Arctic will be prolonged in terms of degree days relative to temperate populations. The correspondingly larger time budget for growth per instar is consistent with the considerably larger size of arctic compared with temperate adults (Table 1).

However, juvenile development also shows a high degree of variation, both between and within life stages. In warm summers many juveniles, especially the younger stages, moult twice (Søvik and Leinaas 2003a), implying that developmental rate decreases with ontogeny. Thus, given a couple of warm summers some specimens may complete their life cycle in 3–4 years. On the other hand, as some juveniles postpone moulting (Søvik and Leinaas 2003a) and females reproduce for 2–3 years (Søvik and Leinaas 2003b), cooler summers may extend the generation time to 6–8 years. This shows that *A. lineatus* has a flexible life cycle well suited to the natural variation in the arctic climate.

Table 1. Average adult sizes (mm) (with ranges when given) of *A. lineatus* from different latitudes.

Place	Females	Males	Authority
Svalbard	0.87 (0.76–1.00)	0.77 (0.69–0.86)	Søvik et al. (2003)
Svalbard	0.763	–	Schubart (1975)
Nordkapp	0.713	–	Schubart (1975)
Scotland	(0.696–0.771)		Colloff (1984)
Limfjorden	0.702	0.636	Schubart (1975)
London	0.670	0.615	Schubart (1975)

Field experiments indicate a sex-dependent developmental rate, being faster for males than females (Søvik and Leinaas 2003a). A male surplus in spring in temperate *A. lineatus* populations (Schubart 1975) indicates earlier male moulting and hence faster male development also at lower latitudes.

Survival and longevity

Survival has only been studied in salt marsh populations (Søvik and Leinaas 2003a). Summer survival is high, with no difference between life stages. Winter survival is also high, but varies between stages, apparently with lowest survival for adults, suggesting a higher juvenile than adult cold tolerance. Where tested, oribatid juveniles have lower supercooling points than conspecific adults (Young and Block 1980; Shimada et al. 1992; Webb and Block 1993; Sugawara et al. 1995; Hansen 2000). The larva-to-adult survival is high (13.3%) (Søvik and Leinaas 2003a), which is consistent with a low lifetime reproductive output (see below). A similar egg-to-adult survival was estimated for the alpine *Oromurcia sudetica* Willmann (9.5%) (Schatz 1983) and the aquatic *Hydrozetes lemnae* Coggi (ca. 8%) (Fernandez and Athias-Binche 1986).

The high summer survival can be explained by an abundant food supply, few or no inter-specific competitors or predators, and stable humidity and temperature conditions in the wet and thermally buffered salt marshes (Søvik et al. 2003). Similarly, the high winter survival suggests that low subzero temperatures are not a major constraint either. Thus, neither abiotic nor biotic factors appear important in the regulation of population density (Søvik and Leinaas 2003a). It is, however, conceivable that density-dependent intra-specific effects may regulate the population.

Adults live for 2–3 years (Figure 3) (Søvik et al. 2003; Søvik and Leinaas 2003a, b). At high experimental temperatures (15–20 °C) longevity decreases, probably because of high reproductive investment and heat stress, whereas at low temperatures (5 °C) *A. lineatus* shows a potential for an extraordinarily high adult longevity with some females surviving 4–5 years (Søvik and Leinaas 2003a, b). The high longevity is probably largely a result of short growing seasons and low temperature budgets. In comparison, most temperate *A. lineatus* adults die in autumn, leaving the tritonymph as the main wintering instar (Bücking et al. 1998). Other temperate species exhibit high adult longevity in the laboratory (365–730 days) (Luxton 1981a).

Survival of arctic adults is influenced by gender, reproductive investment and possibly subzero temperatures. Females survive longer than males (Søvik and Leinaas 2003b), which may explain the seasonally stable female surplus observed in two salt marsh populations (67.3 and 55.3%) (Søvik et al. 2003). Sex-dependent longevity has not been investigated in other oribatid mites, but varying sex ratios in many temperate species (Luxton 1981a), including temperate populations of *A. lineatus* (Schubart 1975), suggest that sex-dependent mortality may be common. A depletion of energy reserves of males during

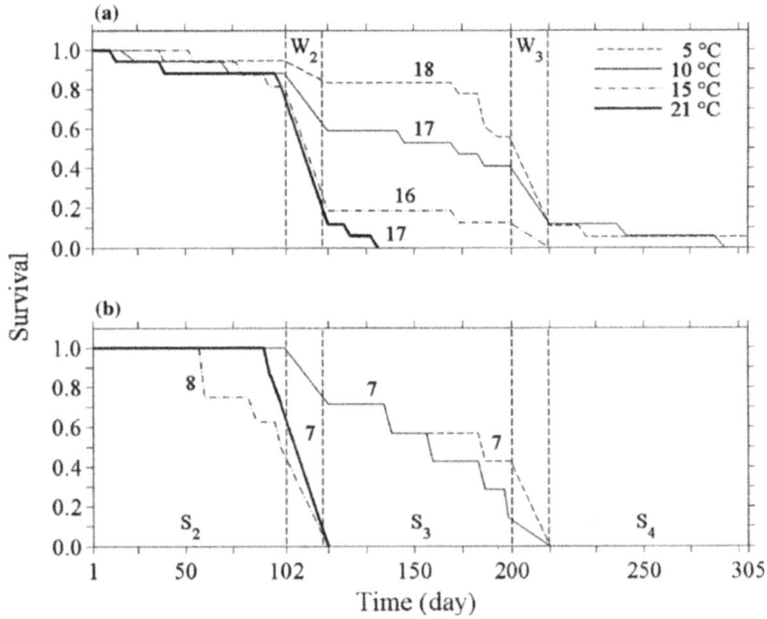

Figure 3. Longevity of females (a) and males (b) through two simulated 'winters' (W_{2-3}) and three simulated 'summers' (S_{2-4}) (with four different temperatures, 5, 10, 15, and 21 °C) in laboratory cultures (after Søvik and Leinaas 2003b). Before start of the experiment the adults had experienced one field summer and one laboratory 'winter'. Time (days after start of experiment) includes only 'summer' days. The 'winters' lasted ca. 4 months. Sizes of the original cohorts are given in the figure.

mating may explain the higher male mortality (Convey 1992). High female reproductive investment during summer results in high mortality the following winter (Søvik and Leinaas 2003b), showing a trade-off between current reproduction and subsequent survival and reproductive performance (Stearns 1997). Finally, a simulated 'winter' in the laboratory with subzero temperatures had a positive effect on newly moulted males, with a higher survival the following summer compared with males reared at constant positive temperatures (Søvik and Leinaas 2003b). No explanation was offered by the authors for this observation.

Reproductive biology

Fertilization in oribatid mites is indirect: males deposit stalked spermatophores, which are subsequently picked up by the females via the genital aperture (Norton 1994). *A. lineatus* probably has indirect fertilization. The unusually long spermatopositor may suggest a closer association of mates

(Norton 1994), however, courtship rituals or mating have never been observed. Furthermore, stalked spermatophores have been found in a few field microcosms (containing only *A. lineatus*), although never in laboratory cultures (G. Søvik, pers. obs.). Females start developing eggs right after moulting and under favourable conditions these mature into larvae the same summer (Søvik et al. 2003). However, no larviposition occurs before the second adult summer (Søvik and Leinaas 2003b). Total embryonic developmental time is estimated to be 2.5–6.0 months, depending on temperature (Søvik and Leinaas 2003b). Females produce eggs continuously and larvae are deposited the whole summer, but periods of high larviposition are typical, often in early summer (Søvik et al. 2003; Søvik and Leinaas 2003b). During normal climatic conditions most females are seasonally iteroparous, larvipositing during two summer seasons. In comparison, the Antarctic *Alaskozetes antarcticus* and *Halozetes belgicae* (Michael) appear to mature a full batch of eggs before oviposition, and after deposition do not seem to produce more eggs (P. Convey, pers. comm.). Søvik and Leinaas (2003b) estimated a lifetime reproductive output of ca. 20 larvae. Instantaneous clutch sizes (Figure 4) (Søvik et al. 2003) are higher than reported for most other temperate and alpine/Antarctic oribatid species (Solhøy 1975; Luxton 1981a; Convey 1994b; Tilrem 1994; Kuriki 1995; Block and

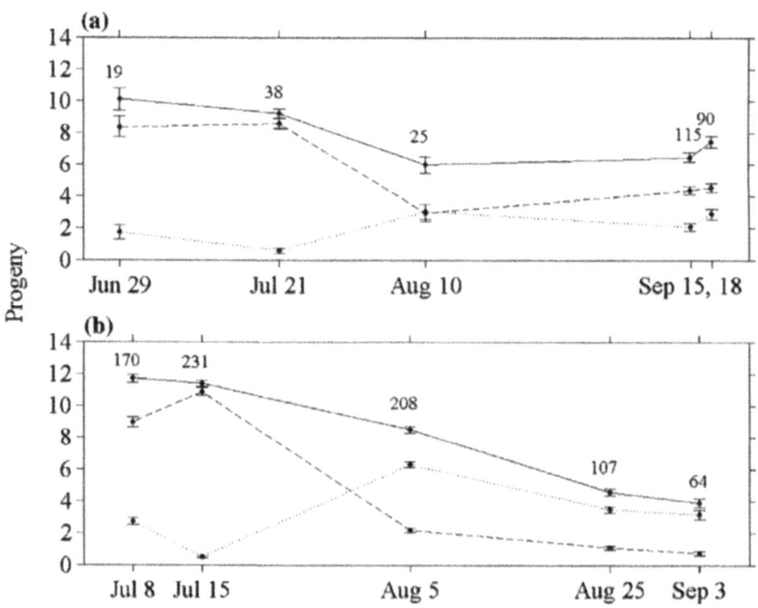

Figure 4. Mean number of eggs (- - -), larvae (...), and total progeny (——) per gravid female (±SE) in a salt marsh population of *A. lineatus* in Adventdalen, West Spitsbergen during two summer seasons, 1997 (a) and 1998 (b) (after Søvik et al. 2003). Sample sizes are given in the figure. Note different scales on *x*-axes.

Convey 1995 and refs. therein; Sugawara et al. 1995; Hubert 2000). Information about species-specific turnover rates is necessary in order to make valid comparisons. Due to larviparity, turnover rate is relatively low in *A. lineatus* and compared with some temperate species (Luxton 1981a; Enami 1992) lifetime reproductive output is low. On the other hand, larviparity reduces egg mortality (Fashing 1975).

Abiotic and biotic influences on reproduction

As mentioned above, embryonic developmental rate depends on microhabitat temperature. This results in considerable annual variation in proportions of gravid females and mean clutch sizes (Figure 4) (Søvik et al. 2003). Temperature dependency of reproduction has also been demonstrated in the laboratory. A constant temperature of 15 °C was shown to be the most favourable for all aspects of reproduction, with highest larviposition rate and shortest pregnancies (Søvik and Leinaas 2003b). However, lifetime reproductive output was equally high at 10 °C. The reproductive performance at 15 °C resembled reproduction in field populations during an unusually warm arctic summer, with high larviposition rate and increased female mortality in late summer (Søvik et al. 2003). Reproductive investment at a mean temperature of 8.7 °C in the field was similar to that obtained at constant 15 °C in the laboratory, which indicates a pronounced positive effect of daily temperature fluctuations, as has also been noted by Lebrun (1977) and Stamou (1989). This shows that in warm summers *A. lineatus* experiences near optimal conditions in its natural habitat and females are able to mature and deposit their whole lifetime reproductive output during one season. Thus, the species reveals a flexible life history by taking advantage of favourable climatic conditions. High reproductive investment at relatively low temperatures may represent a physiological adaptation. Unfortunately, no data on temperature effects on reproduction in temperate *A. lineatus* exist, and without comparative studies it is difficult to evaluate this question. Other temperate oribatid species show highest reproduction at constant temperatures of 20–25 °C (Kaneko 1988; Stamou 1989; Kuriki 1993; Vera 1993). In comparison, arctic *A. lineatus* showed signs of heat stress at 21 °C (Søvik and Leinaas 2003b).

At low experimental temperatures (5 °C) *A. lineatus* deposited eggs with incomplete larval development (Søvik and Leinaas 2003b). Only a few of these eggs completed development, and none of the surviving larvae moulted to protonymphs (Søvik 2003). Søvik suggested that if this mixed-parity mode also occurs in natural habitats, it may represent an alternative reproductive strategy of arctic *A. lineatus* experiencing very low summer temperatures. Ovoviviparity implies long pregnancies lasting several years at low temperatures (Søvik and Leinaas 2003b). By ovipositing, the slow embryonic development would be less dependent on female winter survival, which is lower than for juveniles (Søvik and Leinaas 2003a).

As opposed to low summer temperatures, subzero winter temperatures positively affect all aspects of reproduction the following summer (Søvik and Leinaas 2003b), which suggests that *A. lineatus* is adapted to the highly seasonal arctic environment. In comparison, most temperate adults die before winter (Bücking et al. 1998).

Finally, spatial differences in reproductive biology, despite very similar microclimatic conditions, suggest that the demography of local populations also depends on biotic factors like habitat quality and, possibly, local genetic adaptation (Søvik et al. 2003). Furthermore, demographic patterns not only depend on current conditions, but just as much on the state of the population in early season, and thus, by implication, biotic and abiotic influences the previous summer.

Ecology

Habitat and food

The degree of habitat specialization of *A. lineatus* decreases with increasing latitude (Schuster 1966; Schulte et al. 1975). In the subtropics the species is confined to a particular zone of the marine rocky shore (Schulte et al. 1975; Schulte and Weigmann 1977), whereas in temperate regions it inhabits marine, brackish, and limnetic rocky shores, occurring in crevices, under stones, in algae (*Blidingia* and *Enteromorpha*) and barnacles, and in the lichen zone of the supralittoral (Schulte et al. 1975; Schulte and Weigmann 1977; Pugh and King 1985; Bücking et al. 1998). A preference for brackish habitats has been noted by several authors (Schulte 1979; Colloff 1984 and refs. therein; Bücking et al. 1998). In Iceland it was found at 10 m depth in a freshwater lake (Lindegaard 1992). In cold temperate regions *A. lineatus* also inhabits inland sites in the coastal zone (Sculte et al. 1975), like salt marshes (Luxton 1967a, b) and moss and grasses 200–300 m from the shore (Colloff 1984).

In the Arctic the distribution of *A. lineatus* extends into the eulittoral as well as into fully terrestrial habitats. This can be explained by a temperature-dependent submersion and salinity tolerance, with a larger range of conditions tolerated at low temperatures (Schulte 1978, 1979). The species has been sampled on rocky and sandy beaches, in moist algae near fresh and brackish water, in washed up sea-weed, in salt marshes, on cyanobacteria mats, and in *Enteromorpha* on walls of bird cliffs (Summerhayes and Elton 1923, 1928; Schuster 1966; Behan 1978; Søvik et al. 2003). Furthermore, it has been found in screes, 'fjellmark', and dry tundra, up to 500 m from the shore and at altitudes of 100–300 m above sea level (Summerhayes and Elton 1923, 1928; Schuster 1966; Karppinen 1967; Schulte et al. 1975; Coulson et al. 2000; Søvik et al. 2003). In the more complex habitats, like 'fjellmark' and *Carex* tussocks, *A. lineatus* occurs together with Collembola, Diptera, Hymenoptera, Araneae, and other Acari (Summerhayes and Elton 1923, 1928; Coulson et al. 2000;

G. Søvik, pers. obs.). By contrast, the species was the only arthropod taxon present in the cyanobacteria and *Enteromorpha* habitats, accompanied by Nematoda, Tardigrada, Rotifera, and Enchytraeidae (Summerhayes and Elton 1923; Søvik et al. 2003). The impoverished fauna could be due to high moisture and salinity, or lack of suitable food. Similar low oribatid diversity in polar regions has only been reported from isolated or extreme habitats (e.g., polar desert soil, rocky ground) (McAlpine 1965; Behan 1978; Block and Starý 1996).

Faecal pellets as well as successful moulting and larviposition in cultures reared on pieces of cyanobacteria mats (Søvik and Leinaas 2003a, b) showed that the mites were feeding and thriving on this food. Few terrestrial arthropods are known to graze on cyanobacteria, but the confamilial *Alaskozetes antarcticus*, and an undescribed species in Selenoribatidae and *Fortuynia* in the same superfamily (Ameronothroidea) were observed to do so (Schuster 1979; Burn 1986; Worland and Lukešová 2000). Detailed studies of the diet of arctic *A. lineatus* do not exist. Temperate *A. lineatus* feeds on green algae and unicellular phycobionts, and in the laboratory consumes yeasts, ascomycete fungi, red macroalgae, as well as cyanobacteria (Schulte 1976).

Multi-instar aggregations and migration

Dense multi-instar aggregations comprising active and quiescent mites have been observed on cyanobacteria mats on Spitsbergen, often along stems of *Puccinellia phryganodes* (Søvik et al. 2003), and on the underside of stones near the seashore in Nunavut Territories (Banks 1919) and in screes on Bear Island and Spitsbergen (Summerhayes and Elton 1923; G. Søvik, pers. obs.). Premoult juveniles have been noted to form small aggregations in cultures (Søvik and Leinaas 2003a). This behaviour may be plesiotypic since temperate populations of *A. lineatus*, as well as other littoral *Ameronothrus* species, also form multi-instar aggregations (Schulte et al. 1975; Bücking et al. 1998). Possibly it evolved to prevent dislodgement by waves (Schulte et al. 1975), as indicated by the location of aggregations in crevices and depressions, and the occurrence of younger instars underneath tritonymphs and adults. Similar aggregations in other littoral mites (Astigmata, Prostigmata) (Pugh 1995; Bücking et al. 1998) support this hypothesis. Alternatively, aggregations could aid mate location by ensuring that adults are in close proximity after the final moult, as was hypothesized for aggregations of *Alaskozetes antarcticus* (Block and Convey 1995). A male surplus in the *A. lineatus* aggregations on the cyanobacteria mats indicates that males were attracted to newly moulted females (Søvik et al. 2003). Similar explanations have been suggested for aggregations of the Antarctic mesostigmatid *Gamasellus racovitzai* (Trouessart) (Goddard 1979) and the oribatid *Maudheimia wilsoni* Dalenius (Marshall and Convey 1999). Compact aggregations of quiescent juveniles may also, due to the dark colouration, accelerate development by creating optimum temperatures through increased absorbed insolation (Haq 1982; Block and Convey 1995). The vulnerable pre-moult resting stage will be shortened, and an increased developmental rate may also reduce the number of winterings and thus the concomitant mortality. Group moulting is also known from laboratory cultures of a few oribatid species at lower latitudes (Woodring and Cook 1962; Haq 1982; Ramani and Haq 1988; Honciuc 1996). None of the suggested hypotheses seem to apply to large, temporally circumscribed aggregations of adult females of the parthenogenetic *Hydrozetes lemnae* Coggi in lakes and ponds (Burford 1976), suggesting the existence of other fundamental advantages of aggregations.

As suggested above, males are attracted to aggregations of quiescent juveniles. Alternatively, the skewed sex ratio of the aggregations may be due to migration of females to lower density areas for larviposition (Søvik et al. 2003). This was observed in *G. racovitzai* (Goddard 1979). Both explanations imply a directed migratory behaviour of adults. Little is known about dispersal in oribatid mites, but according to Norton (1994), it probably occurs mainly in the adult instar. Bücking et al. (1998) describe a vertical migration pattern in temperate *A. lineatus* involving both adults and juveniles, where adults move into the uppermost eulittoral in July/August and deutonymphs migrate back to the littoral fringe after 2 months.

Population density and dynamics

Stable stage structure in field populations of arctic *A. lineatus* results from extensive overlapping generations due to the slow juvenile development and females larvipositing throughout summer (Søvik et al. 2003). Inter-annual variations in stage structure are likely to occur. First, increased larviposition in warm summers (Søvik et al. 2003; Søvik and Leinaas 2003b) combined with low larval mortality in winter (Søvik and Leinaas 2003a), will lead to an increased larval density and proportion the following summer. Second, as adult winter survival is influenced by reproduction, adult density will decrease following warm summers (Søvik et al. 2003; Søvik and Leinaas 2003b). Finally, juvenile development depends on temperature and ontogeny (Søvik and Leinaas 2003a), resulting in varying proportions of the different life stages. Stable stage structures are characteristic of many alpine and Antarctic oribatid mites (Solhøy 1975; West 1982; Schatz 1985; Tilrem 1994; Kuriki 1995; Marshall and Convey 1999), whereas in temperate species cohorts often can be distinguished throughout ontogeny (Luxton 1981a, b). The latter pattern is seen in temperate populations of *A. lineatus*, where larviposition is temporally circumscribed in mid-summer and cohorts can be followed to the adult stage (Bücking et al. 1998).

Population density estimates of arctic *A. lineatus* exist only from Spitsbergen. Schuster (1966) stated that it was found ‘with high individual densities’ on rocky and sandy beaches. Similarly, in salt marshes the species occurs in high densities, 1.3–2.7 × 10^4 m^{-2} have been estimated (Søvik et al. 2003; G. Søvik, unpubl. data), but true densities are probably at least twice as high since extraction efficiency was low (Søvik and Leinaas 2002). High densities have also been recorded on cyanobacteria mats (8.1–14.2 cm^{-2} in high density patches) (Søvik et al. 2003). The closely related *Ameronothrus nigrofemoratus* (L. Koch) (Schubart 1975) was found in high abundance (20.6×10^4 m^{-2}) on an identical cyanobacteria surface in East Greenland (Hammer 1944, 1946 published incorrectly as *A. lineatus nigrofemorata* L. Koch (Schubart 1975)). *A. lineatus* is often the only oribatid or arthropod species in such habitats (Søvik et al. 2003). Similarly, Antarctic soil fauna communities often comprise few species, but have high population densities (10^4–10^7 m^{-2}) (Convey 2001). In a temperate *A. lineatus* population, Bücking et al. (1998) recorded 2.0–2.4×10^4 mites m^{-2}, which are of the same size order as the arctic densities, but considerably lower than population densities recorded for other temperate littoral oribatid mites (Luxton 1967a, b; Ernst et al. 1993; Bücking et al. 1998).

The population density remained quite stable throughout an arctic summer in the above mentioned salt marsh and cyanobacteria habitats (Søvik et al. 2003), but no inter-annual data exist. In a 3-year long study of oribatid mites on Svalbard five out of six species differed in abundance among years (Webb et al. 1998).

Conclusions

Arctic and temperate *A. lineatus* populations share many similarities, but also differ in important ways. Shared life-history traits, possibly plesiotypic, include larviparity, multi-instar aggregations, sex-dependent development and survival (resulting in skewed sex ratios), and a flexible life cycle with variable developmental rate and all stages able to overwinter (Bücking et al. 1998). On the other hand, large differences exist in life-cycle duration, adult longevity, and timing of larviposition. Furthermore, a stimulating effect of winter on demography in arctic populations has been demonstrated. Some differences can be explained by the highly different temperature regimes, but arctic populations may also have evolved specific life-history traits adapting them to, and hence allowing them to exploit cold environments. The synchronized moulting, life processes being stimulated by cold exposure, and a high female reproductive investment at relatively low-summer temperatures may represent such evolved traits. The typical oribatid life history, with extended adult survival, low reproductive output and slow development, has been suggested to be pre-adapted to cold environments (Norton 1994; Convey 1996; Behan-Pelletier 1999). This paper suggests that a combination of ancestral and evolved traits contribute to the success of *A. lineatus* in arctic environments. Further studies of temperate *A. lineatus* populations are needed to evaluate the importance of phenotypic plasticity versus microevolutionary adaptations of life-history traits of arctic *A. lineatus*.

Acknowledgements

Barb Eamer processed and provided the SEM-picture. Torstein Solhøy kindly helped with literature search. Two anonymous referees provided helpful comments on earlier versions of this manuscript.

References

Banks N. 1919. The Acarina collected by the Canadian Arctic Expedition, 1913–18. Rep. Can. Arct. Exped. 3: 11–13.

Behan V.M. 1978. Diversity, Distribution and Feeding Habits of North American Arctic Soil Acari. Ph.D. thesis, Dept. of Entomology, McGill University, Canada.

Behan-Pelletier V.M. 1999. Oribatid mite fauna of northern ecosystems: a product of evolutionary adaptations or physiological constraints? In: Needham, Mitchell, Horn and Welbourn (eds), Acarology IX, Vol. 2, Symposia. Ohio Biological Survey, Columbus, Ohio, pp. 87–105.

Block W. and Convey P. 1995. The biology, life cycle and ecophysiology of the Antarctic mite *Alaskozetes antarcticus*. J. Zool. Lond. 236: 431–449.

Block W. and Starý J. 1996. Oribatid mites (Acari: Oribatida) of the maritime antarctic and antarctic peninsula. J. Nat. Hist. 30: 1059–1067.

Bücking J., Ernst H. and Siemer F. 1998. Population dynamics of phytophagous mites inhabiting rocky shores – K-strategists in an extreme environment? In: Ebermann (ed.), Arthropod Biology: Contributions to Morphology, Ecology and Systematics. Biosystematics and Ecology Series 14. Österr. Akad. Wiss, Vienna, pp. 93–143.

Burford D.R. Jr. 1976. Morphology and Life History of *Hydrozetes bushnelli* n. sp. (Oribatei, Hydrozetidae). Ph.D. thesis. Dept. of Environmental, Population, and Organismic Biology, University of Colorado, United States.

Burn A.J. 1986. Feeding rates of the cryptostigmatid mite *Alaskozetes antarcticus* (Michael). Br. Antarct. Surv. Bull. 71: 11–17.

Colloff M.J. 1984. Notes on two lichenophagous oribatid mites from Ailsa Craig (Acari: Cryptostigmata). Glasg. Nat. 20: 451–457.

Convey P. 1992. Seasonal lipid contents of Antarctic micro-arthropods. Exp. Appl. Acarol. 15: 219–231.

Convey P. 1994a. Growth and survival strategy of the Antarctic mite *Alaskozetes antarcticus*. Ecography 17: 97–107.

Convey P. 1994b. Sex ratio, oviposition and early development of the Antarctic oribatid mite *Alaskozetes antarcticus* (Acari: Cryptostigmata) with observations on other oribatids. Pedobiologia 38: 161–168.

Convey P. 1996. The influence of environmental characteristics on life history attributes of antarctic terrestrial biota. Biol. Rev. 71: 191–225.

Convey P. 2001. Antarctic ecosystems. In: Levin (ed.), Encyclopedia of Biodiversity, Vol. 1. Academic Press, San Diego, pp. 171–184.

Coulson S.J., Hodkinson I.D., Strathdee A.T., Block W., Webb N.R., Bale J.S. and Worland M.R. 1995. Thermal environments of arctic soil organisms during winter. Arctic Alpine Res. 27: 364–370.

Coulson S.J., Leinaas H.P., Ims R.A. and Søvik G. 2000. Experimental manipulation of the winter surface ice layer: the effects on a High Arctic soil microarthropod community. Ecography 23: 299–306.

Coulson S.J. and Refseth D. 2004. The terrestrial and freshwater invertebrate fauna of Svalbard (and Jan Mayen). In: Prestrud, Strøm and Goldman (eds), A Catalogue of the Terrestrial and Marine Animals of Svalbard. Skrifter 201. Norwegian Polar Institute, Tromsø.

Danks H.V. 1981. Arctic Arthropods. A Review of Systematics and Ecology with Particular Reference to the North American fauna. Biological Survey project. Entomological Society of Canada, Ottawa.

Danks H.V. 1999. Life cycles in polar arthropods – flexible or programmed? Eur. J. Entomol. 96: 83–102.

Enami Y. 1992. Life history of *Epidamaeus verrucatus* Enami et Fujikawa (Acari: Damaeidae), with morphological description of its immature stage. Edaphologia 48: 23–29.

Ernst H., Siemer F., Bücking J. and Witte H. 1993. Die litorale Milbenzönose auf Uferbefestigungen des Weserästuars in Abhängigkeit von Substrat und Salzgehaltsgradient. Inf. Natursch. Landschaftspfl. 6: 401–416.

Fashing N.J. 1975. Life history and general biology of *Naiadacarus arboricola* Fashing, a mite inhabiting water-filled treeholes (Acarina: Acaridae). J. Nat. Hist. 9: 413–424.

Fernandez P.N.A. and Athias-Binche F. 1986. Analyse démographique d'une population d'*Hydrozetes lemnae* Coggi, Acarien Oribate inféodé à la lentille d'eau *Lemna gibba* L. en Argentine. 1. Méthodes et techniques, démographie d' *H. lemnae* comparaisons avec d'autres Oribates. Zool. Jahrb. Abt. Syst. Oekol. Geogr. Tiere 113: 213–228.

Gilyarov M.S. 1975. A Key to the Soil-inhabiting Mites. Sarcoptiformes. Nauka Publ, Moscow. (in Russian).

Goddard D.G. 1979. Biological observations on the free-living mites of Signy Island in the maritime Antarctic. Br. Antarct. Surv. Bull. 49: 181–205.

Hammer M. 1944. Studies on the Oribatids and Collemboles of Greenland. Meddelelser om Grønland 141: 4–210.

Hammer M. 1946. The Zoology of East Greenland. Oribatids. Meddelelser om Grønland 122: 2–39.

Hansen M.P. 2000. Seasonal Variation in Tolerance of Cold and Drought in *Ameronothrus lapponicus* (Acari: Oribatida) from Finse, Norway. Cand. Scient. thesis, Dept. of Zoology, University of Bergen, Norway.

Haq M.A. 1982. Pheromonal regulation of aggregation and moulting in *Archegozetes longisetosus* (Acari, Oribatei). Calicut Univ. Res. J., Special Conference Number May 1982: 19.

Haq M.A., Sheela K. and Neena P. 1991. Ovoviviparity in oribatid mites. In: Veeresh, Rajagopal and Viraktamath (eds), Advances in Management and Conservation of Soil Fauna. Oxford & Publishing Co. Pvt. Ltd, New Delhi, pp. 763–767.

Honciuc V. 1996. Laboratory studies of the behaviour and life cycle of *Archegozetes longisetosus* Aoki 1965 (Oribatida). In: Mitchell, Horn, Needham and Welbourn (eds), Acarology IX. Proceedings. Ohio Biological Survey, Columbus, Ohio, pp. 637–640.

Hubert J. 2000. Seasonal changes of abundance, sex ratio and egg production of *Scheloribates laevigatus* (Acari: Oribatida) in the soil of a meadow in the Czech Republic. Acta Soc. Zool. Bohem. 64: 37–56.

Kaneko N. 1988. Life history of *Oppiella nova* (Oudemans) (Oribatei) in cool temperate forest soils in Japan. Acarologia 29: 215–221.

Karppinen E. 1967. Notes on the arthropod fauna of Spitsbergen I. 2. Data on the oribatids (Acari) of Spitsbergen. Ann. Ent. Fenn. 33: 18–26.

Karppinen E. and Krivolutsky D.A. 1982. List of oribatid mites (Acarina, Oribatei) of northern palaearctic region. I. Europe. Acta Entomol. Fenn. 41: 1–18.

Kuriki G. 1993. Reproductive process and developmental rate of *Trhypochthoniellus setosus* Willmann (Oribatida). J. Acarol. Soc. Jpn. 2: 7–13. (in Japanese with English abstract).

Kuriki G. 1995. Life cycle of *Trhypochthoniellus setosus* Willmann (Acari: Trhypochthoniidae) in a *Sphagnum* moor at Yachidaira, Northeast Japan. J. Acarol. Soc. Jpn. 4: 113–122.

Lebrun P. 1977. Comparaison des effets des temperatures constantes ou variables sur la durée de développement de *Damaeus onustus* (Acarina: Oribatei). Acarologia 19: 136–143.

Lindegaard C. 1992. Zoobenthos ecology of Thingvallavatn: vertical distribution, abundance, population dynamics and production. Oikos 64: 257–304.

Luxton M. 1967a. The zonation of saltmarsh Acarina. Pedobiologia 7: 55–66.

Luxton M. 1967b. The ecology of saltmarsh Acarina. J. Anim. Ecol. 36: 257–277.

Luxton M. 1981a. Studies on the oribatid mites of a Danish beech wood soil. IV. Developmental biology. Pedobiologia 21: 312–340.

Luxton M. 1981b. Studies on the oribatid mites of a Danish beech wood soil. VI. Seasonal population changes. Pedobiologia 21: 387–409.

Marshall D.J. and Convey P. 1999. Compact aggregation and life-history strategy in a continental Antarctic mite. In: Bruin, van der Geest and Sabelis (eds), Ecology and Evolution of the Acari. Kluwer Academic Publishers, Dordrecht, pp. 557–567.

McAlpine J.F. 1965. Insects and related terrestrial invertebrates of Ellef Ringnes Island. Arctic 18: 73–103.

Mitchell M.J. 1977. Population dynamics of oribatid mites (Acari, Cryptostigmata) in an aspen woodland soil. Pedobiologia 17: 305–319.

Norton R.A. 1994. Evolutionary aspects of oribatid mite life histories and consequences for the origin of the Astigmata. In: Houck (ed.), Mites. Ecological and Evolutionary Analyses of Life-history Patterns. Chapman & Hall, New York, pp. 99–135.

Pugh P.J.A. 1995. Air-breathing littoral mites of sub-Antarctic South Georgia. J. Zool. Lond. 236: 649–666.

Pugh P.J.A. and King P.E. 1985. The vertical distribution of the British intertidal Acari – the non halacarid fauna (Arachnida: Acari). J. Zool. Lond. (A) 207: 21–33.

Pugh P.J.A. and King P.E. 1986. Seasonality in some British intertidal Acari. J. Nat. Hist. 20: 653–666.

Ramani N. and Haq M.A. 1988. Developmental studies of *Uracrobates indicus* (Acari: Oribatei) inhabiting *Mangifera indica*. In: Channabasavanna and Viraktamath (eds), Progress in Acarology, Vol. 1. Oxford and IBH Publishing Co., New Dehli, pp. 483–489.

Schatz H. 1983. Überlebensrate von *Oromurcia sudetica* Willmann (Acari, Oribatei) von einer alpinen Wiese Tirols (Obergurgl, Zentralalpen). Zool. Jb. Syst. 110: 97–109.

Schatz H. 1985. The life cycle of an alpine oribatid mite, *Oromurcia sudetica* Willmann. Acarologia 26: 95–100.

Schubart H. 1970. *Ameronothrus schusteri* n. sp., eine neue Oribatide von der Küste Jugoslawiens. Senckenbergiana Biol. 51: 425–432.

Schubart H. 1975. Morphologische Grundlagen für die Klarung der Verwandt-schaftsbeziehungen innerhalb der Milbenfamilie Ameronothridae (Acari, Oribatei). Zoologica 123: 23–65.

Schulte G. 1976. Zur Nahrungsbiologie der terrestrischen und marinen Milbenfamilie Ameronothridae (Acari, Oribatei). Pedobiologia 16: 332–352.

Schulte G. 1978. Die Küstenbindung terrestrischer Arthropoden und ihre Bedeutung für den Wandel des Ökosystems "marines Felslitoral" in unterschiedlichen geographischen Breiten. Mitt. dtsch. Ges. allg. angew. Ent. 1: 211–219.

Schulte G. 1979. Brackwasser-Submergenz und latitudinale Emergenz küstenbewohnender Milben. In: Piffl (ed.), Proceedings of the 4th International Congress of Acarology. Akademiai Kiado, Budapest, pp. 61–67.

Schulte G., Schuster R. and Schubart H. 1975. Zur Verbreitung und Ökologie der Ameronothriden (Acari, Oribatei) in terrestrischen, limnischen und marinen Lebensräumen. Veröff. Inst. Meeresforsch. Bremerh. 15: 359–385.

Schulte G. and Weigmann G. 1977. The evolution of the family Ameronothridae (Acari: Oribatei). II. Ecological aspects. Acarologia 19: 167–173.

Schuster R. 1966. Hornmilben (Oribatei) als Bewohner des marinen Litorals. Veröff. Inst. Meeresforsch. Bremerh. Sonderband II: 319–327.

Schuster R. 1979. Soil mites in the marine environment. In: Rodriguez (ed.), Recent Advances in Acarology. Academic Press, New York, pp. 593–602.

Schuster R. 1988. Transoceanic distribution of air-breathing littoral mites. In: Channabasavanna and Viraktamath (eds), Progress in Acarology, Vol. 1. Oxford and IBH Publishing Co., New Delhi, pp. 355–362.

Shimada K., Pan C. and Ohyama Y. 1992. Variation in summer cold-hardiness of the Antarctic oribatid mite *Alaskozetes antarcticus* from contrasting habitats on King George Island. Polar Biol. 12: 701–706.

Solhøy T. 1975. Dynamics of oribatei populations on Hardangervidda. In: Wielgolaski (ed.), Ecological Studies 17. Fennoscandian Tundra Ecosystems Part 2. Animals and Systems Analysis. Springer-Verlag, Berlin, pp. 60–65.

Soma K. 1990. Studies on the life history of *Phthiracarus japonicus* Aoki (Acarina: Phthiracaridae) in a creeping pine (*Pinus pumila* Regel) shrub. Edaphologia 43: 25–30. (in Japanese with English abstract).

Søvik G. 2003. Observations on ovoviviparity and mixed-parity mode in arctic populations of *Ameronothrus lineatus* (Acari, Oribatida). Acarologia 43: 393–398.

Søvik G. and Leinaas H.P. 2002. Variation in extraction efficiency between juvenile and adult oribatid mites: *Ameronothrus lineatus* (Oribatida, Acari) in a Macfadyen high-gradient canister extractor. Pedobiologia 46: 34–41.

Søvik G., Leinaas H.P., Ims R.A. and Solhøy T. 2003. Population dynamics and life history of the oribatid mite *Ameronothrus lineatus* (Acari, Oribatida) on the high arctic archipelago of Svalbard. Pedobiologia 47: 257–271.

Søvik G. and Leinaas H.P. 2003a. Long life cycle and high adult survival in an arctic population of the mite *Ameronothrus lineatus* (Acari, Oribatida) from Svalbard. Polar Biol. 26: 500–508.

Søvik G. and Leinaas H.P. 2003b. Adult survival and reproduction in an arctic mite, *Ameronothrus lineatus* (Acari, Oribatida): effects of temperature and winter cold. Can. J. Zool. 81: 1579–1588.

Stamou G.P. 1989. Studies on the effect of temperature on the demographic parameters of *Achipteria holomonensis* (Acari, Oribatida). Acarologia 30: 171–180.

Stearns S.C. 1997. The Evolution of Life Histories. Oxford University Press, Oxford.

Strenzke K. 1955. Oribates (Acariens). In: Arachnides, Strenzke, de Lesse and Denis (eds), Microfaune du sol de l'eqe Groenland. I. Hermann & Cie, Éditeurs, Paris, pp. 14–64.

Sugawara H., Ohyama Y. and Higashi S. 1995. Distribution and temperature tolerance of the Antarctic free-living mite *Antarcticola meyeri* (Acari, Cryptostigmata). Polar Biol. 15: 1–8.

Summerhayes V.S. and Elton C.S. 1923. Contributions to the ecology of Spitsbergen and Bear Island. J. Ecol. 11: 214–286.

Summerhayes V.S. and Elton C.S. 1928. Further contributions to the ecology of Spitsbergen. J. Ecol. 16: 193–268.

Thörell T. 1871. Om Arachnider från Spitsbergen och Beeren-Eiland. Öfvers. Kongl. Vet.-Akad. Förh. 28: 683–702. (in Swedish).

Tilrem L. 1994. Life History Traits in two Oribatid mites (*Ameronothrus lapponicus* Dalenius and *Phauloppia* sp.) in an Extreme High Mountain Habitat. Cand. Sci. thesis. Dept. of Zoology, University of Bergen, Norway.

Vera H. 1993. Demographic variation in two forest populations of oribatid mites. Pedobiologia 37: 95–106.

Walter D.E. and Proctor H.C. 1999. Mites. Ecology, Evolution and Behaviour. CABI Publishing, Wallingford.

Webb N.R. and Block W. 1993. Aspects of cold hardiness in *Steganacarus magnus* (Acari: Cryptostigmata). Exp. Appl. Acarol. 17: 741–748.

Webb N.R., Coulson S.J., Hodkinson I.D., Block W., Bale J.S. and Strathdee A.T. 1998. The effects of experimental temperature elevation on populations of cryptostigmatic mites in high Arctic soils. Pebobiologia 42: 298–308.

Weigmann G. and Schulte G. 1975. *Ameronothrus schubarti* n. sp. aus dem marinen Litoral Kaliforniens (Arachnida: Acari: Oribatei). Senckenbergiana Biol. 56: 133–143.

Weigmann G. and Schulte G. 1977. The evolution of the family Ameronothridae (Acari: Oribatei) I. Morphological aspects. Acarologia 19(1): 161–166.

West C. 1982. Life histories of three species of sub-antarctic oribatid mite. Pedobiologia 23: 59–67.

Woodring J.P. and Cook E.F. 1962. The biology of *Ceratozetes cisalpinus* Berlese, *Scheloribates laevigatus* Koch, and *Oppia neerlandica* Oudemans (Oribatei), with a description of all stages. Acarologia 4: 101–137.

Worland M.R. and Lukešová A. 2000. The effect of feeding on specific soil algae on the cold-hardiness of two Antarctic micro-arthropods (*Alaskozetes antarcticus* and *Cryptopygus antarcticus*). Polar Biol. 23: 766–774.

Young S.R. and Block W. 1980. Experimental studies on the cold tolerance of *Alaskozetes antarcticus*. J. Insect Physiol. 26: 189–200.

Experimental and Applied Acarology **34:** 21–35, 2004.

Latitudinal variation in habitat specificity of ameronothrid mites (Oribatida)

DAVID J. MARSHALL[1,*] and PETER CONVEY[2]
*[1]Department of Biology, Universiti Brunei Darussalam, Jalan Tungku Link, Gadong BE 1410, Brunei Darussalam; [2]British Antarctic Survey, Natural Environment Research Council, High Cross, Madingley Road, Cambridge CB3 0ET, UK; *Author for correspondence (e-mail: marshall@fos.ubd.edu.bn; phone: +673-2-249001 ext. 1385; fax: +673-2-461502)*

Key words: Acari, Antarctic, Arctic, Biogeographical distribution, Niche specificity, Southern hemisphere

Abstract. Ameronothroid mites, including Ameronothridae, Fortuyniidae and Selenoribatidae, are unique among the Oribatida through having a global distribution from the tropics to the poles, and occupying a diversity of habitats including terrestrial, marine and freshwater. Their ecological diversification is of considerable interest from both the perspective of evolution over geological timescales, and the detail of the underlying processes. Given their widespread global distribution, it seems likely that historical global events (tectonic and climatic) have played a fundamental role in their ecological diversification. Previous studies of sub-Antarctic island arthropods have generated considerable circumstantial evidence in support of glaciation being a primary factor influencing ecological patterns: lower habitat specificity and weaker interspecific interactions are associated with more recent (postglacial) vegetated terrestrial biotopes, as compared to the older epilithic and littoral biotopes (which are assumed to have been present, *albeit* reduced in extent, during Neogene glacial maxima). Here, we use ameronothrid mites as a case study to examine the extent to which the above island scenario generalizes globally across latitudes affected by glaciation. We show that, unlike congeners or even conspecifics at lower latitudes in each hemisphere which are restricted to marine environments, the species found at higher latitudes (especially *Alaskozetes antarcticus*, *Ameronothrus dubinini*, *Ameronothrus lineatus*, and *Halozetes belgicae*) show greater affinity for terrestrial environments. They show a transition or expansion of habitat use (from marine-influenced to terrestrial habitats) implicit with a lower degree of habitat specificity, in relation to increasing latitude. We contend that the terrestrial environment at higher latitudes in both hemispheres has been colonized by these ameronothrid mite species following the various glaciation events, facilitated by a lack of competition experienced in their low diversity communities, in a manner which represents a larger scale demonstration of the processes described on sub-Antarctic islands.

Introduction

Among the most significant factors influencing terrestrial biotic patterns over geological time is climate change and, particularly, its effect on the glaciation of land masses. This has resulted in prominent large-scale species extinction in the polar regions, as evidenced today by low terrestrial biodiversity, particularly in the Antarctic (Convey 2001). Because climate change may also cause shifts in latitudinal distributions of species, there has been an increasing interest in studying the latitudinal variation of taxonomic, ecological and ecophysiological attributes (Rohde 1992; Crame 1992, 1993; Chown and Clarke 2000), in order to better understand and monitor the effects of present-day climate change. Climate induced reductions in species ranges, for example, have implications for population extinctions and even species extinctions. The present study aims to identify glaciation effects and latitudinal variation in habitat utilization of ameronothrid mites, and thereby to improve understanding of how species distributions may respond to present-day climate change.

Ameronothroid mites are unusual among the Oribatida in their habitat diversity, especially the adoption by many species of an aquatic existence. Representatives are found in freshwater streams, high altitudinal ephemeral pools, salt marshes and mangroves, as well as on intertidal rocky shores (Behan-Pelletier 1997; Norton et al. 1997; Procheś and Marshall 2001). In addition, the group shows an exceptionally broad geographical distribution, and is represented from the polar to the equatorial regions. For the marine-associated taxa, the tropical and warm temperate species belong to Selenoribatidae and Fortuyniidae whereas the Ameronothridae occur towards the sub-polar and polar regions. Different origins have been ascribed to these geographically separated groups, with species interactions hypothesized to have driven the marine origination of the tropical selenoribatid and fortuyniid mites, and glaciation thought to be the cause of the marine existence by the polar ameronothrid mites (Wallwork 1965; Schuster 1966; Procheś 2001; Procheś and Marshall 2001, 2002).

Although a variety of mite species occur in the marine environment, only halacarid mites (Prostigmata) manage complete transition, with their capacity for aquatic respiration, a submerged lifestyle, and occupation of the deep sea benthos (these have been classified as primary marine mites; Procheś and Marshall 2001). Secondary marine mites, on the other hand, which are confined to intertidal zones and air-breathe, comprise the Ameronothroidea (Oribatida), Hyadesioidea (Astigmata), and representatives of the Mesostigmata and Prostigmata. Although ameronothroid and hyadesioid mites interact trophically with other marine biota, littoral-inhabiting mesostigmatid and prostigmatid mites retain strong terrestrial affinities, with single species either occurring in both environments or being represented by terrestrial congeners (Procheś and Marshall 2001).

Although some species of ameronothrid mites are typically marine, others occupy terrestrial habitats, and this as well as their polar distributions, make them particularly suitable for studying the influence of glaciation on the evolution of habitat utilization in marine and terrestrial environments.

Several investigations have considered habitat utilization in relation to latitude for the northern hemisphere ameronothrid genus, *Ameronothrus* (Schulte 1975; Schulte et al. 1975; Schulte and Weigmann 1977; Søvik et al. 2003; Søvik 2004). However, the evolutionary contexts of these studies are limited, as are those of earlier work concerning geographical distributions of Antarctic arthropods (Wallwork 1967, 1973), which do not address underlying processes for the observed patterns (including the possibility of postglacial habitat expansion). More recently, our knowledge of the biogeography and ecology of peri-Antarctic mites (including the regions conventionally described as sub- and maritime Antarctic) has increased considerably (Starý 1995; Starý and Block 1995, 1996, 1998; Block and Starý 1996; Convey and Quintana 1997; Convey and Smith 1997; Convey et al. 2000a; Pugh and Convey 2000). This has allowed for more realistic comparison of northern and southern hemisphere ameronothrid mite distributions. Furthermore, we now also have a better understanding of habitat utilization by sub-Antarctic microarthropods, including ameronothrid mites (Marshall et al. 1999, 2003; Mercer et al. 2000; Gabriel et al. 2001; Barendse 2002; Chown et al. 2002; Marshall and Chown 2002). All of these studies show large ecological differences among marine littoral, supralittoral, and vegetated and epilithic terrestrial biotopes, in terms of arthropod habitat specificity and interspecific interactions, and propose that such differences relate to glaciation effects.

As a consequence of local extinction during glaciation, contemporary vegetated terrestrial communities on many of the sub-Antarctic islands are thought to comprise colonising species originating from older epilithic biotopes (marine or terrestrial) unaffected by glaciation (Chown 1990, 1994; Barendse et al. 2002). Given the apparent diversity and abundance of resources in terrestrial environments, the component species show little habitat (or niche) specificity in comparison with their intertidal, supralittoral or epilithic counterparts, which probably experienced intense competition when the islands were glaciated. The ecological consequences of glaciation, by the reduction or elimination of available terrestrial habitat, should be as applicable globally as they are to the sub-Antarctic islands, and investigation into this formed the basis of the present study. Here, we examine the degree of postglacial habitat expansion shown by marine-associated ameronothrid mite lineages across their global range, and specifically address two questions: whether postglacial habitat expansion (through utilization of terrestrial habitats) is more evident in higher latitudinal species and the related question of whether habitat specialization (through a marine or supralittoral existence) occurs predominantly in lower latitudinal species.

Methods

The available literature was reviewed to compile data on the ecological and latitudinal distributions of known Ameronothridae mite species. The focus was on the polar, sub-polar and cold temperate regions, therefore the typically tropical and temperate families (Selenoribatidae, Fortuyniidae) were excluded from the analysis. The ameronothrid genera, *Antarcticola* Wallwork 1967, *Aquanothrus* Engelbrecht 1975, *Capillibates* Hammer 1966, *Chudalupia* Wallwork 1981, and *Pseudantarcticola* Balogh 1970, were also excluded either because their ecological and geographical distributions are poorly known or their present-day distributions indicate that they are not relevant to the hypotheses under consideration. For example, one genus has an ancient low latitudinal and high altitudinal origination (*Pseudantarcticola* from New Guinea). Other genera remain as relict populations in inland, ephemeral freshwater bodies (*Aquanothrus* and *Chudalupia*; Engelbrecht 1975; Wallwork 1981; Norton et al. 1997), whereas the high latitudinal *Antarcticola* shows no association with the marine environment (Wallwork 1967, 1973) and, unlike other maritime Antarctic and sub-Antarctic ameronothrid mites, its distribution on South Georgia, Kerguelen and Continental Antarctica suggests a Gondwanan origination (see Marshall and Coetzee 2000). Our study thus concerned the following four ameronothrid genera: *Alaskozetes* Hammer 1966, *Ameronothrus* Berlese 1896, *Halozetes* Berlese 1917, and *Podacarus* Grandjean 1955, which have affiliations to marine-influenced or Southern Ocean island habitats, and representation at relatively high latitudes in either the northern or southern hemispheres. Within these genera, undescribed species were omitted from the analyses, as was an unconfirmed record of *Ameronothrus bilineatus* in southern Africa, which can only be explained as the result of human translocation (Weigmann 1975).

Most studies on the ameronothrid taxa concern their systematics. Even though information on habitat utilization is valuable in understanding the origins of mites, efforts have only recently been made to provide objective ecological information. This is particularly so for the peri-Antarctic systems (e.g. Block and Convey 1995, for *Alaskozetes antarcticus*), and many habitat and niche descriptions for species in this region remain fragmentary. Although frequent reference in the general literature to a plant or biotic condition (detritus, penguin skull, penguin rookery) may indicate incidental local distribution, it often informs little about preferential habitat use or any evolutionary selection of habitat. Recent studies on Marion Island have shown that habitat specificity of microarthropods can be assessed by means of quantitative ecological investigation, while also highlighting the need to distinguish between species in their typical habitat and so-called 'tourist species' which have been transposed from adjacent habitat (Mercer et al. 2000; Gabriel et al. 2001; Barendse et al. 2002; Chown et al. 2002).

Although the ameronothrid mites of the northern hemisphere have been the subject of considerable ecological investigation, there remains some inconsistency between studies (e.g. contrasting habitats given for mites in Schulte 1975 and Schulte and Weigmann 1977). Additionally, direct comparison between the hemispheres is limited by differences in the kinds on habitat and the ways they have been defined in studies for each hemisphere. Although there is a clear distinction between marine, supralittoral and terrestrial habitats for the southern hemisphere (Gabriel et al. 2001; Barendse et al. 2002; Marshall and Chown 2002; Marshall et al. 2003), this is not always so for the northern hemisphere, especially with reference to the supralittoral and terrestrial habitats. Some confusion seemingly relates to the poorly defined sediment-based marine systems (e.g. salt-marshes) in the northern hemisphere, which are not present (or at least not significant) in the peri-Antarctic. In the present study, the supralittoral zone of the peri-Antarctic is defined as the marine-influenced epilithic biotope directly above the intertidal zone, which is characterized by the presence of the lichens *Turgidosculum* sp. and *Caloplaca* sp. and sometimes the alga *Prasiola* (Mercer et al. 2000; Barendse et al. 2002; Chown et al. 2002; Marshall and Chown 2002; Marshall et al. 2003). This zone represents a transitional zone between the land and the sea and is thought to have been a glacial refuge, which together with its high arthropod habitat specificity necessitates separate consideration from the marine and terrestrial zones (Mercer et al. 2000; Barendse et al. 2002).

Given the difficulty for direct comparison between the hemispheres, the central hypothesis is examined separately for each hemisphere. For the northern hemisphere, latitudinal distribution records were based on a selection of review articles, particularly Schulte (1975), Schulte et al. (1975), Schulte and Weigmann (1977), Hammer and Wallwork (1979), Pugh and King (1985), and Procheś and Marshall (2001). Similarly, so as to not repeat the information given in numerous studies from the southern hemisphere, records were extracted from the summaries and reviews of Wallwork (1967), Luxton (1990), Pugh (1993), and Starý and Block (1998). Latitudinal ranges for the southern hemisphere were based on the latitudinal positions of the islands on which the mite species occur. Relevant papers concerning habitat utilization of the peri-Antarctic ameronothrid mites included Block and Convey (1995), Marshall et al. (1999), Chown et al. (2002), and Marshall and Chown (2002).

Results and discussion

Northern hemisphere fauna

The geographical distributions of the northern hemisphere ameronothrid mite genus *Ameronothrus*, show a clear relationship between habitat occupation and latitude (Table 4 in Schulte et al. 1975; Schulte and Weigmann 1977; Table 1). At lower latitudes, only the intertidal zone is occupied, including either rocky-shores or the soft substrata associated with saltmarshes (*Ameronothrus schusteri*, *A. schubarti*, *A. bilineatus*, *A. schneideri*, and *A. marinus*; Schulte and Weigmann 1977; Table 1). At the higher peri-Arctic latitudes there is a shift towards occupation of the intermediate supralittoral zone, and ultimately terrestrial habitats. Although a single species in the northern part of the distribution of *Ameronothrus* is entirely terrestrial (*A. dubinini* (= *A. lapponicus*)), 3 other species occurring at peri-Arctic latitudes also maintain marine associations across much of their latitudinal distribution (*A. nigrofemoratus*, *A. lineatus*, *A. maculatus*; Table 1). These species have been recorded in limnetic and coastal terrestrial habitats in addition to the supralittoral and intertidal zones, but become more stenotopic and restricted to marine-influenced environments at their southerly distributions (see *A. nigrofemoratus* and particularly *A. lineatus*; Table 1; Schulte et al. 1975; Schulte 1976; Schulte and Weigmann 1977; Søvik et al. 2003; Søvik 2004).

Table 1. Northern hemisphere ameronothrid mite (*Ameronothrus* spp.) habitats and regional distributions.

Species	Habitat	Region
Ameronothrus schusteri (Schubart 1970)	L	Mediterranean; Atlantic
Ameronothrus schubarti (Weigmann and Schulte 1975)	L	California
Ameronothrus bilineatus (Michael 1888)	L	Eastern Atlantic-Boreal; Mediterranean
Ameronothrus schneideri (Oudemans 1903)	L	Eastern Atlantic- Boreal; Mediterranean; Tropical Western Atlantic
Ameronothrus marinus (Banks 1896)	L	Eastern Atlantic-Boreal; Mediterranean
Ameronothrus maculatus (Michael 1882)	SL/T	Arctic; Eastern Atlantic-Boreal; Mediterranean
Ameronothrus nigrofemoratus (L. Koch 1879)	L/SL/T	Arctic; Eastern Atlantic-Boreal; Eastern Pacific Boreal
Ameronothrus lineatus (Thörell 1871)	L/SL/T	Arctic; Eastern Atlantic-Boreal; Eastern Pacific Boreal
Ameronothrus dubinini (Sitnikova 1975) (= *lapponicus* Dalenius 1963)	T	Arctic; sub-Arctic

Habitat abbreviations are L: marine intertidal, SL: supralittoral zone, T: terrestrial.
Table compiled by reference to, primarily, Schulte (1975), Schulte and Weigmann (1977), Schulte et al. (1975), Hammer and Wallwork (1979), Pugh and King (1985), Procheś and Marshall (2001), Søvik and Leinaas (2002, 2003), and Søvik et al. (2003).

The presence of *Alaskozetes coriaceus* in the northern hemisphere and *A. antarcticus* in the southern hemisphere is currently the only bipolar generic distribution within the ameronothrid mites, as previous reference to bipolarity for *Ameronothrus* (Schulte and Weigmann 1977), requires confirmation. However, as the 2 *Alaskozetes* species differ markedly in ecological and geographical distributions and are grouped on the basis of a tentative taxonomic character (shape of the anterior notogaster), the accuracy of their phylogenetic relationship is best considered with reservation until molecular evidence is available.

Southern hemisphere fauna

Of the 16 species included in the peri-Antarctic ameronothrid genera (*Alaskozetes*, *Halozetes* and *Podacarus*), three (*A. antarcticus*, *H. belgicae* and *H. marinus*) show conspicuously broad latitudinal distributions extending over 25–35° of latitude (Table 2). These species occur at the southernmost latitudes occupied by ameronothrid mites on the Antarctic Peninsula (Wallwork 1967; Pugh 1993; Convey et al. 2000b); although a single specimen of *A. antarcticus* has been obtained during an air sampling programme in Continental Antarctica (Wallwork 1967, 1973), no living material has been reported or collected

Table 2. Southern hemisphere ameronothrid mites.

Species	Habitat	Latitudinal range	Localities
Alaskozetes antarcticus (Michael 1903)	SL/T	45–70	F,SG,SO,SE,AP,MP,C,K,H,M
Halozetes marinus (Lohmann 1907)	L	35–65	F,SG,SO,MP,C,K,H,M,CA,AS
Halozetes belgicae (Michael 1903)	SL/T	35–70	SG,SO,SE,AP,MP,C,K,H,M,CA,AS
Halozetes marionensis (Engelbrecht 1974)	L	40–50	MP,G
Halozetes fulvus Engelbrecht 1975	T	45–50	MP
Halozetes crozetensis (Richters 1907)	T	35–55	F,C,K,H,M,CA,AS
Halozetes intermedius (Wallwork 1963)	L	45–55	H,M,K
Halozetes impeditus (Niedbala 1986)	U	60–65	SE
Halozetes littoralis (Wallwork 1970)	U	50–55	SG,M
Halozetes necrophagus (Wallwork 1974)	U	60–65	AP
Halozetes plumosus (Wallwork 1966)	U	50–55	CA
Halozetes bathamae (Luxton 1985)	L	45–50	NZ
Halozetes macquariensis (Dalenius and Wilson 1958)	T	50–55	M,CA
Halozetes otagoensis (Hammer 1966)	SL	40–50	NZ
Halozetes capensis (Coetzee and Marshall 2003)	L	30–35	SA
Podacarus auberti (Grandjean 1955)	T	45–55	SG,MP,C,K,H,M

Habitat, latitudinal ranges and localities of *Alaskozetes*, *Halozetes* and *Podacarus* species. Habitats are L: intertidal, SL: occurring within the epilithic supralittoral zone, dominated by *Prasiola* and lichens (especially *Turgidosculum* and *Caloplaca*), T: terrestrial. 'U' indicates conflicting or unclear habitat descriptions. Subspecies and undescribed species are not included in the analysis. Latitudinal ranges are accurate to 5° bands.
Localities and islands: Falklands (F); South Georgia (SG); South Orkney (SO); South Shetland (SE); Antarctic Peninsula (AP); Gough (G); Marion and Prince Edward (MP); Crozet (C); Kerguelen (K); Heard (H); Amsterdam and St. Paul (AS); South Africa (SA); Macquarie (M); Campbell (CA); New Zealand (NZ). Prominent references used to compile the table were Wallwork (1967); Luxton (1990); Pugh (1993); Starý and Block (1998), and Marshall and Chown (2002).

and the species is not considered to be resident there. Additionally, the above-mentioned 3 species are the most widespread and abundant (Richard et al. 1994; Block and Convey 1995; Convey and Smith 1997; Convey et al. 2000b), and show the highest degree of morphological variation among the southern hemisphere ameronothrid mites, with the numerous described sub-species

Table 3. Habitat utilization of ameronothrid mites on the peri-Antarctic Heard and Marion Islands (see Marshall et al. 1999; Chown et al. 2002; Marshall and Chown 2002).

Habitat	Marion island	Heard island
Terrestrial	*Halozetes fulvus* (Engelbrecht) *Podacarus auberti* (Grandjean)	*Halozetes crozetensis* Richters *Podacarus auberti auberti* Grandjean *Podacarus auberti occidentalis* (Wallwork)
Supralittoral	*Alaskozetes antarcticus intermedius* (Michael)	*Alaskozetes antarcticus antarcticus* (Michael) *Alaskozetes antarcticus grandjeani* (Dalenius)
	Halozetes belgicae mickii (Coetzee)	*Halozetes belgicae belgicae* (Michael) *Halozetes belgicae brevipilis* (Wallwork)
Upper-shore marine	*Halozetes marionensis* (Engelbrecht)	*Halozetes intermedius* (Wallwork)
Lower-shore marine	*Halozetes marinus devilliersi* (Engelbrecht)	*Halozetes marinus marinus* (Lohmann)

The supralittoral habitat comprises lichens *Caloplaca* and *Turgidosculum* (previously known as *Mastodia*), the upper-shore comprises the lichen, *Verrucaria*, and the mid to low-shore comprises seaweeds. The terrestrial biotope comprises a variety of habitat types including vegetative and epilithic habitats. Although sub-species are presented here, their taxonomic distinctness remains contentious (see Convey et al. 2000a,b).

suggesting evolution through geographical isolation. However, the frequent occurrence of 2 sub-species in similar habitat at the same location leads to questions around the validity of these sub-species (see Table 3). Most of the remaining southern hemisphere species occur within 45–55°S latitudes, with an odd exception provided by the distribution of the recently described *Halozetes capensis* from southern Africa (30–35°S), which is most likely linked to the cold Benguela current off the west coast of the region (Coetzee and Marshall 2003).

Unlike the northern hemisphere, there is no clear relationship between habitat occupation and latitudinal distribution for the southern hemisphere ameronothrid mites, at least when habitat is coarsely defined as terrestrial, supralittoral or marine (Table 2). Although the northernmost *Halozetes* representatives on New Zealand and in southern Africa (*Halozetes bathamae*, *H. otagoensis*, and *H. capensis*) are limited to marine habitats, 3 species occurring on the mid-latitudinal, peri-Antarctic islands (*Halozetes fulvus*, *H. crozetensis* and *Podacarus auberti*) are confined to terrestrial habitats. The latter situation may arise through the independent establishment of environmental gradients on the islands and differential glaciation effects, relating to oceanic currents and geomorphological features (altitude), in addition to latitudinal climatic effects. Alternatively, latitudinally separated islands may possess similar microhabitat conditions, and thus show similarities in aspects of their arthropod ecology. This is exemplified by identical habitat separation and specificity of the representative ameronothrid mites on Marion Island (46°S), and the more southerly located Heard Island (53°S; south of the oceanic Polar Frontal Zone, and currently 80% glaciated) (Table 3; Starý and Block 1998).

In the upper-shore marine and epilithic supralittoral zones of Marion and Heard islands, the respective mite species show a high degree of niche specificity and are restricted to particular lichen types, whereas the terrestrial mite species exhibit low niche specificity in their occupation of various plant types and other habitats in epilithic and vegetative biotopes (though *Podacarus auberti* clearly favours *Poa cookii* which grows in the seaward areas; Table 3; Mercer et al. 2000; Barendse et al. 2002). This pattern provides insights into both the evolutionary origin of these mite taxa as well as historical aspects of the habitats. It suggests that niche-specific speciation preceded an episode of within-niche evolution across the region, and, that the low niche specificity of the contemporary terrestrial taxa relates to their more recent colonization, probably from older, ice-free, epilithic biotopes. The repeat information for these islands confirms differences in habitat specificity of the various species, but, the extent to which this occurs widely across the South Indian Ocean Province is obscured by imprecise habitat descriptions for Kerguelen and Crozet (Wallwork 1963, 1965; Travé 1974, 1976).

Rather than focussing on specific examples of species restricted to single islands, some clarification is obtained by considering species more generally distributed across latitude, such as *Halozetes belgicae* and *Alaskozetes antarcticus*. Whereas both these species occupy very narrow epilithic, supralittoral niches at their more northern peri-Antarctic island distributions (Table 3), they commonly occur (or have occurred in the past) in terrestrial habitats removed from direct marine influence at their southern latitudinal distributions. At Signy Island (60°S), *A. antarcticus* is found in a variety of terrestrial habitats below 200 m asl and occurs predominantly above the supralittoral (Goddard 1979), a pattern that appears to be repeated throughout the maritime Antarctic at latitudes between 60 and 70°S (Tilbrook 1973; Schenker and Block 1986; Richard et al. 1994; Block and Convey 1995; Convey and Smith 1997; Convey et al. 2000b). In these studies, it has been recorded from a wide range of vegetated habitats (including algae, mosses, lichens, grass), as well as rocks, soils and organic debris associated with vertebrate colonies. In contrast, on the peri-Antarctic South Georgia, Marion and Heard islands, its distribution is much more tightly restricted to supralittoral sites along the coast (PC pers. obs., Marshall and Chown 2002).

Although Goddard (1979) reported that *H. belgicae* was prominent in the supralittoral zone, but also rarely found in freshwater pools and lakes at Signy, in subsequent surveys (refs. above) the species has been located in a similarly wide range of habitats as *A. antarcticus*. The habitat preferences of the 2 species are not identical, with *H. belgicae* not showing the same abundance and association as *A. antarcticus* with vertebrate breeding and moulting sites, rather being more associated with epilithic crustose lichens. Both species show considerable ability to survive when submerged in fresh or seawater, although there is no suggestion of these being permanent habitat (Goddard 1979; Pugh and MacAllister, 1994). Palaeolimnological studies of the organic remains in lake sediments on Signy island have identified both species (Jones et al. 2000), with their abundance in sediment layers correlating both with climatic reconstructions and also interpretations of the different habitat requirements of the 2 species and changes in the balance of these habitat types (Hodgson and Convey submitted).

In contrast to the supralittoral northern peri-Antarctic island species (*Alaskozetes antarcticus* and *Halozetes belgicae*), the terrestrial species seemingly show little variation in their habitat use across their relatively limited latitudinal distribution ranges. For example, *Podacarus auberti* favours *Poa* vegetation on both Marion Island and the much more southerly South Georgia Island (Wallwork 1966; Marshall et al. 1999).

Habitat and niche specificity

The ecological distribution patterns of ameronothrid mites in the southern and northern hemispheres raise several questions about control mechanisms, in particular, (1) what controls niche specificity at lower latitudes (especially in the case of the peri-Antarctic mites), and (2) what controls the variation in this towards higher latitudes. Although no studies have measured competition among ameronothrid mites, there is seemingly an abundance of food in peri-Antarctic habitats and competition or resource partitioning is thought to be an unlikely cause of spatial separation of microarthropods (Usher and Booth 1984; Burn and Lister 1988; Convey 1996a,b; Barendse et al. 2002). Nevertheless, the restriction of *Halozetes belgicae* and *Alaskozetes antarcticus* towards supralittoral zones in the sub-Antarctic is suggestive of some interspecific interaction, particularly as at least one other similarly-sized species is found in the adjacent terrestrial habitats on these islands (*Halozetes fulvus*, *Halozetes crozetensis* and *Podacarus auberti*; Marshall et al. 1999; Marshall and Chown 2002; Table 3).

Possession of a greater ecophysiological ability to withstand more extreme conditions may underlie the more southerly distribution of *Halozetes belgicae* and *Alaskozetes antarcticus*. However, confirmation of this will require detailed ecophysiological studies across other species in the group that have yet to be attempted. Indeed, whereas *A. antarcticus* clearly possesses considerable stress tolerance ability, it can also be described as showing little specialization to the Antarctic environment with respect to this ability as well as to feeding and growth, as these abilities all appear to be ancestral within temperate ameronothrid mites (Norton 1994; Block and Convey 1995). As a generalist (herbivore and detritivore), it has been found to feed on crustose lichen, alga (especially *Prasiola crispa*) and organic debris (Goddard 1979, 1982). Towards the southern limits of their distributions, both *H. belgicae* and *A. antarcticus* may utilize these generalist feeding abilities to exploit a variety of habitats within a terrestrial environment devoid of real competitors. The same generalization may also underlie the ability of northern circumpolar *Ameronothrus* spp. (Table 1) to occupy terrestrial habitats (cf. Søvik 2004). It is clear from the foregoing that the roles of feeding and niche specialization require clarification through detailed studies, including the description of genetic differences among populations (and across latitudes), digestive enzyme energetics, the absorptive component of the food, the gut faunal composition, and other physiological attributes.

Habitat specificity in the context of glacial history

Glaciation has been suggested to be a key factor in determining the present-day ecological patterns of sub-Antarctic island arthropods. Lower habitat specificity and weaker interspecific interactions are associated with the postglacial, edaphic habitats that emerged as a result of ice-cap recession, as compared to the older epilithic and littoral habitats, which are assumed to have been present as refuges on most sub-Antarctic islands during the Neogene glacial maxima (Chown 1994; Barendse et al. 2002; Chown et al. 2002). Our study suggests that such glaciation-driven ecological effects may operate over a global scale, and are seemingly the cause of the observed pattern of decreasing habitat specificity of ameronothrid mites with increasing latitude, in both the southern and northern hemispheres. In summary, ameronothrid mites occurring at high latitudes in each hemisphere (especially of the genera *Alaskozetes*, *Ameronothrus*, *Halozetes*) show greater affinity for terrestrial habitats, unlike their congeners or even the same species occurring at lower latitudes. This is achieved either by an expansion of habitat utilization (from marine-influenced to terrestrial) or by becoming entirely terrestrial and thus, by implication, exhibiting a lower degree of niche specificity.

An alternative hypothesis to that proposed here, that the high latitudinal (and by implication terrestrial) ameronothrid species for the regions under consideration are ancestral (Schulte and Weigmann 1977), and presumably survived the effects of glaciation in terrestrial epilithic refugia, fails to explain many ecological and geographical distribution patterns of the peri-Antarctic mites. For example, it does not explain, (1) why the terrestrial island species (*Halozetes fulvus*, *H. crozetensis* and *Podacarus auberti*) do not occur in the higher latitudinal zones and, (2) why *Halozetes belgicae* and *Alaskozetes antarcticus* show high habitat specificity and are restricted from the terrestrial habitats of the sub-Antarctic islands. Furthermore, this hypothesis appears to not have considered habitat availability in relation to glaciation history for the northern hemisphere. For the southern hemisphere, current models for the Pleistocene glacial maxima indicate both considerably greater depth of ice sheets than now present, along with greater geographical expansion with ice sheets extending to the edge of the continental shelf throughout the Antarctic Peninsula region (Larter and Vanneste 1995; Convey 2003). These models raise questions concerning the identification of: (1) possible locations for glacial refuge sites for terrestrial biota during the glacial maxima and (2) the mechanism(s) by which Antarctic terrestrial invertebrates (including arthropods) have successfully dispersed, to effectively occupy terrestrial 'islands' across a wide geographical range (see Convey 2003 for discussion), and support suggestions of postglacial recolonization from lower southern hemisphere latitudes.

Our proposal that the terrestrial environment at higher latitudes contains postglacial recolonising ameronothrid mites contradicts some phylogenetic suggestions. Schulte and Weigmann (1977) suggested terrestrial ancestry for the genus *Ameronothrus*, considering that the high latitudinal *A. dubinini* possesses more plesiomorphic characters than its congeners. However, Wallwork (1965) previously argued that other high latitudinal species (*Alaskozetes antarcticus*, *Halozetes belgicae*, *Podacarus auberti*) are derivative forms (based on the expression of sexual dimorphism), and that the typically marine *H. marinus* and *H. intermedius*, are primitive. Although all marine-associated ameronothrid lineages must ultimately have had an earlier terrestrial ancestry (see Procheś and Marshall 2001), possibly linked with the *Aquanothrus*-like mites when considering their antiquity (based on distributions in similar inland geological formations in South Africa, Australia, and North America; Engelbrecht 1975; Wallwork 1981; Norton et al. 1997), a more complete understanding requires thorough morphological and/or molecular phylogenetic investigation of the superfamily.

Acknowledgements

D.J.M. is indebted to Steven Chown for ongoing support, and intellectual and logistical contributions to his Antarctic mite research. The manuscript was much improved by comments from two anonymous referees.

References

Barendse J., Mercer R.D., Marshall D.J. and Chown S.L. 2002. Habitat specificity of mites on Marion Island. Environ. Entomol. 31: 612–625.

Behan-Pelletier V.M. 1997. The semiaquatic genus *Tegeocranellus* (Acari: Oribatida: Ameronothroidea) of North and Central America. Can. Entomol. 129: 537–577.

Block W. and Convey P. 1995. The biology, life cycle and ecophysiology of the Antarctic mite *Alaskozetes antarcticus* (Michael). J. Zool. 236: 431–449.

Block W. and Starý J. 1996. Oribatid mites (Acari: Oribatida) of the maritime Antarctic and Antarctic Peninsula. J. Nat. Hist. 30: 1059–1067.

Burn A.J. and Lister A. 1988. Activity patterns in an Antarctic arthropod community. Br. Antarct. Surv. Bull. 78: 43–48.

Chown S.L. 1990. Possible effects of Quaternary climate change on the composition of insect communities of the South Indian Ocean Province Islands. S. Afr. J. Sci. 86: 386–391.

Chown S.L. 1994. Historical ecology of sub-Antarctic weevils: patterns and processes on isolated islands. J. Nat. Hist. 28: 411–433.

Chown S.L. and Clarke A. 2000. Stress and the geographic distribution of marine and terrestrial animals. In: Storey K.B. and Storey J. (eds), Cell and Molecular Responses to Stress. Elsevier, Amsterdam, pp. 41–54.

Chown S.L., McGeoch M. and Marshall D.J. 2002. Diversity and conservation of invertebrates at the Prince Edward Islands. Afr. Entomol. 10: 67–82.

Coetzee L. and Marshall D.J. 2003. A new *Halozetes* species (Acari, Oribatida, Ameronothroidae) from the marine littoral of southern Africa. Afr. Zool. 38: 327–331.

Convey P. 1996a. The influence of environmental characteristics on the life history attributes of Antarctic terrestrial biota. Biol. Rev. 71: 191–225.

Convey P. 1996b. Overwintering strategies of terrestrial invertebrates from Antarctica – the significance of flexibility in extremely seasonal environments. Eur. J. Entomol. 93: 489–505.

Convey P. 2001. Antarctic Ecosystems. In: Levin S.A. (ed.), Encyclopedia of Biodiversity, Vol. 1. Academic Press, San Diego, pp. 171–184.

Convey P. 2003. Maritime Antarctic climate change: signals from terrestrial biology. In: Domack P., Burnett A., Leventer A., Convey P., Kirby M. and Bindschadler R. (eds), Antarctic Peninsula Climate Variability: A Historical and Palaeoenvironmental Perspective, Antarctic Research Series, Vol. 79. American Geophysical Union, pp. 145–158.

Convey P., Greenslade P. and Pugh P.J.A. 2000a. Terrestrial fauna of the South Sandwich Islands. J. Nat. Hist. 34: 597–609.

Convey P. and Quintana R.D. 1997. The terrestrial arthropod fauna of Cierva Point SSSI, Danco Coast, northern Antarctic Peninsula. Eur. J. Soil Biol. 33: 19–29.

Convey P. and Smith R.I.L. 1997. The terrestrial arthropod fauna and its habitats in northern Marguerite Bay and Alexander Island, maritime Antarctic. Antarct. Sci. 9: 12–26.

Convey P., Smith R.I.L., Peat H.J. and Pugh P.J.A. 2000b. The terrestrial biota of Charcot Island, eastern Bellingshausen Sea, Antarctica an example of extreme isolation. Antarct. Sci. 12: 406–413.

Crame J.A. 1992. Evolutionary history in the polar regions. Hist. Biol. 6: 37–60.

Crame J.A. 1993. Latitudinal range fluctuations in the marine realm through geological time. Trends Ecol. Evol. 8: 162–166.

Engelbrecht C.M. 1975. New ameronothroid (Oribatei, Acari) taxa from the Republic of South Africa and the islands of Gough and Marion. Navors. Nas. Mus. Bloemfontein. 3: 53–85.

Gabriel A.G.A., Chown S.L., Barendse J., Marshall D.J., Mercer R.D., Pugh P.J.A. and Smith V.R. 2001. Biological invasions of Southern Ocean islands: the Collembola of Marion Island as a test of generalities. Ecography 24: 421–430.

Goddard D.G. 1979. The Signy Island terrestrial reference sites: XI. Population studies on the Acari. Br. Antarct. Surv. Bull. 48: 71–92.

Goddard D.G. 1982. Feeding biology of free-living Acari at Signy Island, South Orkney Islands. Br. Antarct. Surv. Bull. 51: 290–293.

Hammer M. and Wallwork J.A. 1979. A review of the world distribution of oribatid mites (Acari: Cryptostigmata) in relation to continental drift. Biol. Skr. Dan. Vid. Selsk. 22: 1–31.

Hodgson D.A. and Convey P. A 7000 year record of the oribatid mite communities on a maritime-Antarctic island: responses to climate change. Arct. Antarct. Alp. Res., (in press).

Jones V.J., Hodgson D.A. and Chepstow-Lusty A. 2000. Palaeolimnological evidence for marked Holocene environmental changes on Signy Island, Antarctica. Holocene 10: 43–60.

Larter R.D. and Vanneste L.E. 1995. Relict subglacial deltas on the Antarctic Peninsula outer shelf. Geology 23: 33–36.

Luxton M. 1990. The marine littoral mites of the New Zealand region. J. R. Soc. New Zeal. 20: 367–418.

Marshall D.J. and Chown S.L. 2002. The acarine fauna of Heard Island. Polar Biol. 25: 688–695.

Marshall D.J. and Coetzee L. 2000. Historical biogeography and ecology of the continental Antarctic mite genus, *Maudheimia* (Acari; Oribatida): evidence for a Gondwanan origin and Pliocene-Pleistocene speciation. Zool. J. Linn. Soc. 129: 111–128.

Marshall D.J., Gremmen N.J.M., Coetzee L., OConnor B.M., Pugh P.J.A., Theron P.D. and Ueckermann E.A. 1999. New records of Acari from the sub-Antarctic Prince Edward Islands. Polar Biol. 21: 84–89.

Marshall D.J., OConnor B.M. and Pugh P.J.A. 2003. *Algophagus* mites (Astigmata: Algophagidae) from the sub-antarctic Prince Edward Islands: habitat-related morphology and taxonomic descriptions. J. Zool. 259: 31–47.

Mercer R.D., Chown S.L. and Marshall D.J. 2000. Mite and insect zonation on a Marion Island rocky shore: a quantitative approach. Polar Biol. 23: 766–774.

Norton R.A. 1994. Evolutionary aspects of oribatid mite life histories and consequences for the origin of the Astigmata. In: Houck M.A. (ed.), Mites, Ecological and Evolutionary Analyses of Life-history Patterns. Chapman & Hall, London, pp. 99–135.

Norton R.A., Graham T.B. and Alberti G. 1997. A rotifer-eating ameronothroid (Acari: Ameronothroidae) mite from ephemeral pools on the Colorado Plateau. In: Mitchell R., Horn D.J., Needham G.R. and Welbourn W.C. (eds), Acarology IX, Proceedings (IXth International Congress of Acarology). Ohio. Biol. Survey, Columbus, pp. 539–542.

Procheś Ś. 2001. Back to the sea: secondary marine organisms from a biogeographical perspective. Biol. J. Linn. Soc. 74: 197–203.

Procheś Ś. and Marshall D.J. 2001. Global distribution patterns of non-halacarid marine intertidal mites: implications for their origins in marine habitats. J. Biogeogr. 28: 47–58.

Procheś Ś. and Marshall D.J. 2002. Diversity and biogeography of southern African intertidal Acari. J. Biogeogr. 29: 1201–1216.

Pugh P.J.A. 1993. A synonymic catalogue of the Acari from Antarctica, the sub-Antarctic Islands and the Southern Ocean. J. Nat. Hist. 27: 323–421.

Pugh P.J.A. and Convey P. 2000. Scotia Arc Acari: antiquity and origin. Zool. J. Linn. Soc. 130: 309–328.

Pugh P.J.A. and King P.E. 1985. The vertical distribution of the British intertidal Acari – the non halacarid fauna (Arachnida: Acari). J. Zool 207: 21–33.

Pugh P.J.A. and MacAllister H.E. 1994. Acari of the supralittoral zone on sub-Antarctic South Georgia. Pedobiologia 38: 552–565.

Richard K.J., Convey P. and Block W. 1994. The terrestrial arthropod fauna of the Byers Peninsula, South Shetland Islands. Polar Biol. 14: 371–379.

Rohde K. 1992. Latitudinal gradients in species-diversity – the search for the primary cause. Oikos 65: 514–527.

Schenker R. and Block W. 1986. Micro-arthropod activity in three contrasting terrestrial habitats on Signy Island, Maritime Antarctic. Br. Antarct. Surv. Bull. 71: 31–43.

Schulte G. 1975. Holoarktische Artareale der Ameronothridae (Acari, Oribatei). Veröff. Inst. Meeresforsch. Bremerh. 15: 339–357.

Schulte G. 1976. Zur Nahrungsbiologie der terrestrischen und marinen Milbenfamilie Ameronothridae (Acari, Oribatei). Pedobiologia 16: 332–352.

Schulte G., Schuster R. and Schubart H. 1975. Zur Verbreitung und Ökologie der Ameronothriden (Acari, Oribatei) in terrestrischen, limnischen und marinen Lebensräumen. Veröff. Inst. Meeresforsch. Bremerh. 15: 359–385.

Schulte G. and Weigmann G. 1977. The evolution of the family Ameronothridae (Acari : Oribatei). II. Ecological aspects. Acarologia. 19: 167–173.

Schuster R. 1966. Hornmilben (Oribatei) als Bewohner des marinen Litorals. Veröff. Inst. Meeresforsch. Bremerh. Sonderband II: 319–327.

Søvik G. 2004. The biology and life history of arctic populations of the littoral mite *Ameronothrus lineatus* (Acari, Oribatida). Exp. Appl. Acarol. 34: 3–20.

Søvik G. and Leinaas H.P. 2002. Variation in extraction efficiency between juveniles and adult oribatid mites: *Ameronothrus lineatus* (Oribatida, Acari) in a Macfadyen high-gradient canister extractor. Pediobiologia 46: 34–41.

Søvik G. and Leinaas H.P. 2003. Long life cycle and high adult survival in an arctic population of the mite *Ameronothrus lineatus* (Acari, Oribatida) from Svalbard. Polar Biol. 26: 500–508.

Søvik G., Leinaas H.P., Ims R.A. and Solhøy T. 2003. Population dynamics and life history of the oribatid mite *Ameronothrus lineatus* (Acari, Oribatida) on the high arctic archipelago of Svalbard. Pedobiologia 47: 257–271.

Starý J. 1995. Oribatid mites (Acari: Oribatida) of Beauchêne Island, Falklands, South Atlantic. J. Nat. Hist. 29: 1461–1467.

Starý J. and Block W. 1995. Oribatid mites (Acari: Oribatida) of South Georgia, South Atlantic. J. Nat. Hist. 29: 1469–1481.

Starý J. and Block W. 1996. Oribatid mites (Acari: Oribatida) of the Falkland Islands, South Atlantic and their zoogeographical relationships. J. Nat. Hist. 30: 523–535.

Starý J. and Block W. 1998. Distribution and biogeography of oribatid mites (Acari : Oribatida) in Antarctica, the sub-Antarctic islands and nearby land areas. J. Nat. Hist. 32: 861–894.

Tilbrook P.J. 1973 Terrestrial arthropod ecology at Signy Island, South Orkney Islands. PhD thesis, University of London.

Travé J. 1974. Observations preliminaries sur les oribates de l'archipel de Kerguelen. In: Piffl E. (ed.), Proceedings of the 4th International Congress of Acarology. Akademiai Kiado, Budapest, pp. 39–45.

Travé J. 1976. Recherches sur les Micrarthropodes terrestres de l'archipel des Kerguelen Données quantitatives – Analyse de deux groupes d'Acariens Oribatida et Acaridida. Rev. Ecol. Biol. Sol. 13: 55–67.

Usher M.B. and Booth R.G. 1984. Arthropod communities in a Maritime Antarctic moss-turf habitat: three-dimensional distribution of mites and Collembola. J. Anim. Ecol. 53: 427–441.

Wallwork J.A. 1963. The Oribatei (Acari) of Macquarie Island. Pac. Insects 5: 721–769.

Wallwork J.A. 1965. Some cryptostigmatid mites (Acari: Cryptostigmata) from Crozet Islands. Pac. Insects 14: 27–37.

Wallwork J.A. 1966. Some Cryptostigmata (Acari) from South Georgia. Br. Antarct. Surv. Bull. 9: 1–20.

Wallwork J.A. 1967. Cryptostigmata (Oribatid mites). In: Gressitt J.L. (eds), Entomology of Antarctica (Antarctic Research Series 10). In: Gressit J.L.(ed.), Entomology of Antarctica (Antarctic Research Series 10). Washington DC American Geiphysical Union, pp. 105–122.

Wallwork J.A. 1973. Zoogeography of some terrestrial microarthropoda in Antarctica. Biol. Rev. 48: 233–2259.

Wallwork J.A. 1981. A new aquatic oribatid mite from western Australia (Acari: Cryptostigmata: Ameronothridae). Acarologia 22: 333–339.

Weigmann G. 1975. Vorkommen von *Ameronothrus* (Acari, Oribatei) im Litoral Südafrikas. Veröff. Inst. Meeresforsch. Bremerh. 15: 65–67.

Experimental and Applied Acarology **34:** 37–58, 2004.

Geographical and ecological distribution of marine halacarid genera and species (Acari: Halacaridae)

ILSE BARTSCH

Forschungsinstitut Senckenberg, Notkestr. 85, 22607 Hamburg, Germany; (e-mail: bartsch@meeresforschung.de; phone: +49-40-89981876; fax: +49-40-89981871)

Key words: Biogeography, Dispersal, Endemisms, Halacaridae, Latitudinal and vertical diversity

Abstract. At the end of 2002, the number of marine halacarid species was 1018, that of genera 51. A single genus, *Copidognathus* contains 33% of all species (336). Eleven genera are monotypic. Geographical provinces with a large number of species are the tropical western Pacific, temperate northeastern Atlantic, temperate southeastern Pacific, and Mediterranean-Black Sea. Most records of halacarid species are from temperate and tropical areas; 10% of species are known from polar zones. On a generic level, 29 genera are recorded from tropical and temperate but not from polar provinces, five genera are restricted to the tropics, and none to polar regions. The majority (920 species or 90%) of all species live in the upper 200 m. Records of genera with exclusively algivorous or brackish/fresh water species are bound to littoral habitats; all the other genera occur in more than one depth zone. Arenicolous genera, though most abundant in the littoral zone, have representatives in the bathyal. Four marine genera (*Copidognathus*, *Halacarellus*, *Isobactrus*, *Lohmannella*) have representatives in coastal fresh water, and three genera, *Acarothrix*, *Caspihalacarus* and *Peregrinacarus*, are predominantly inhabitants of diluted brackish and fresh water. None of the free-living halacarid genera of the world's oceans appears to be endemic to one geographical province.

Introduction

The first species and genera of marine mites were diagnosed and described from the northern Atlantic Ocean at about the middle of the 19th century (Johnston 1836; Gosse 1855a, b; Hodge 1860; 1863; Packard 1871). At present, records of more than 1000 marine species in more than 50 genera exist from all over the world. Within the past decade, from 1993 to 2002, 244 new species and seven new genera have been diagnosed. The geographical areas best known with regard to their halacarid faunas are the shores and shallows of the northern Atlantic and the adjacent basins Baltic, Mediterranean and Black Sea. Extensive sampling has also been done along the Pacific coast of southern South America, the coast of southwestern and northeastern Australia, and Antarctica, whereas from other regions, from the South American Atlantic coast and the African shores, knowledge is fragmentary. The majority of collections have been taken in the littoral, in the tidal and shelf zone, but halacarids are present also in the bathyal and abyssal. Future collections in all regions are expected to result in a large number of new records of species and genera. This paper presents a compilation of published records and the geographical and ecological range of genera and species, followed by a discussion of possible latidudinal and vertical diversity gradients, endemisms and dispersal.

Material

The number of species compiled in the tables are those reported in literature until the end of 2002. Records thought to be based on misidentification, contamination or mislabelling are omitted. Subspecies with characters distinctly separating them from the nominate species are treated as separate species whereas subspecies with weak diagnostic differences are included in the nominate form. Nomina nuda have been left out of consideration. Freshwater representatives of marine genera are included in the tables and attributed to the geographical region of the adjacent sea. Genera known from both marine and limnic environments or exclusively from nearshore fresh water are included, whereas genera mainly spread in continental fresh water are excluded, though they often are found in coastal brackish water. The genus *Troglohalacarus*, with *T. dentipes* Viets 1937, is omitted; according to Bartsch (1989a) the species may belong to the genus *Halacarellus*. The genera are listed in an alphabetical order and not arranged within subfamilies. The classification and definition of halacarid subfamilies are in need of revision. The given geographical boundaries (Figure 1) largely follow Hedgpeth (1963), Lüning (1985) and Briggs (1995).

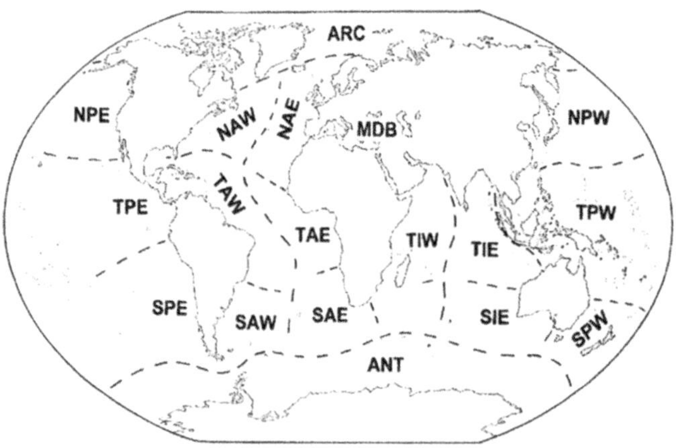

Figure 1. Geographical provinces, modified from Hedgpeth (1963), Lüning (1985) and Briggs (1995). ANT, Antarctica and sub-Antarctica; ARC, Arctic; MDB, Mediterranean and Pontocaspian area; NAE, temperate northeastern Atlantic; NAW, temperate northwestern Atlantic; NPE, temperate northeastern Pacific, NPW, temperate northwestern Pacific; SAE, temperate southeastern Atlantic; SAW, temperate southwestern Atlantic; SIE, temperate southeastern Indian Ocean; SPE, temperate southeastern Pacific; SPW, temperate southwestern Pacific; TAE, tropical eastern Atlantic; TAW, tropical western Atlantic; TIE, tropical eastern Indian Ocean; TIW, tropical western Indian Ocean; TPE, tropical eastern Pacific; TPW, tropical western Pacific.

In a biogeographical aspect, the Indian Ocean is in general treated as a single province, but to differentiate between the African and Asian-Australian coasts, the faunas of eastern and western part of the Indian Ocean are listed separately (TIE and TIW). The Indian Ocean is unique in that surface flow changes direction depending on the seasonal monsoon (Duxbury et al. 2000). The tropical eastern Indian Ocean is separated from the tropical western Pacific. Durban, South Africa, is included in the tropical zone, though the greatest discontinuities in the marine flora and temperature is further north, near the Mozambique border (Bolton et al. 2004). The Mediterranean and the Ponto-Caspian area, with the Black, Caspian and Aral Seas, are united although temperature and salinity are different in these basins. The 180th degree of longitude is taken to separate between the northeastern and northwestern Pacific, accordingly the outer Aleutian Islands are within the northwestern Pacific area.

When compiling vertical distribution, intermediate depths are included; e.g., a species with records from the shelf and a deep sea basin is expected to be present also in the bathyal. The 1000 m depth line was chosen to distinguish between the slope zone and the semi-abyssal and abyssal zone (cf. Haedrich et al. 1975; Gage et al. 1984), instead of the 2000 m line as proposed by, e.g., Sanders and Hessler (1969), Gage and Tyler (1991) and Zezina (1997). Species without any hint to their depth distribution are excluded from the data set. With regard to habitat aspects, only presence (+) or absence (−) of records of a given genus are documented. The randomness of distribution was tested with a Chi-square test.

General data on biology and ecology of halacarid mites

Halacarid mites are benthic, present from the upper littoral fringe to deep sea trenches. Adults are within a size range of 140–2000 μm, most species are less than 500 μm in length. Halacarids live on and between organisms, often within the small spaces between algal scrubs, colonial organisms and gills, as well as in the interstitia of sandy deposits. Halacarids cannot swim but can at least reduce their sinking rate by spreading the legs. Marine mites have not been found on offshore drifting algae (Bartsch 1982a; Lohmann 1889), the scattered records are from nearshore rafts (Viets 1936; Lohmann 1889). The halacarid fauna on algae cast ashore is sparse compared with algae dredged or collected by diving.

Knowledge of population structure and reproduction of halacarids is available primarily from the boreal northern Atlantic and adjacent basins (Lohmann 1889, 1893; Makkaveeva 1966; Straarup 1968; Kirchner 1969; Bartsch 1972; Pugh and King 1986; Siemer 1996). In the majority of populations studied, there is a single generation a year, either with the egg deposition almost simultaneously or with a prolonged period of egg development, with a female carrying only one or two eggs at the same time. At least some of the warm-temperate and tropical halacarid species are expected to have a 1-year life span, too (Bartsch 1992b, 1996b). Data from polar regions are lacking. In general, reproduction and development of gonads, eggs and juvenile stages are influenced by the temperature (Kirchner 1969; Bartsch 1972; Siemer 1996).

Halacarids run through one larval and one to three nymphal stages before the final moult to the adult. Occasionally, nymphal stages are slightly more widespread than adults are, but in general, juveniles and adults of free-living halacarids live in almost the same substratum and horizon. However, both the substratum and the horizon inhabited by a given species may differ considerably from site to site (pers. observ.). Specially adapted migration or dispersal stages are not known. Information on the life cycles of parasitic halacarids is not available. The majority of the marine halacarid species are bisexuals, the only parthenogenetic species presently known is *Isobactrus setosus* (Lohmann 1889). Fecundity of free-living species seems to be low, mostly less than 20 eggs per female lifetime. In contrast, the parasitic species *Enterohalacarus minutipalpus* Viets 1938 has an unusual large number of eggs ('eine auffallend große Anzahl von kleinen Eiern') (Viets 1938); in a slide re-examined there are more than 200 eggs per female (Bartsch, pers. observ.).

Results

Number of genera and species

Table 1 includes 1018 halacarid species in 51 genera. *Copidognathus* is the genus most rich in species, the 336 species comprising 33% of all marine halacarid species. The genus is eurytopic. Although feeding has not been observed, many may feed on bacteria and fungi. The genus *Rhombognathus* currently includes 93 species, equalling 9% of all halacarid species. As demonstrated by the green colour of the gut, phytophagy is the most important way of feeding. The two genera *Agauopsis* and *Halacarus* both include 74 species or about 7% of all halacarid species. Representatives of both genera are expected to be predacious, as e.g. documented by Krantz (1970) and MacQuitty (1984). The other genera include less than 50 species each.

Eleven out of the 51 genera are represented by a single species each. Eight of the genera (*Arenihalacarus*, *Corallihalacarus*, *Enterohalacarus*, *Phacacarus*, *Spongihalacarus*, *Thalassophthirius*, *Werthelloides*, *Xenohalacarus*) are known only from their type locality, where they were present with one to about two dozen individuals (Abé 1991; Bartsch 1986b, 1988b, 1992a; Otto 1999, 2000a, b; Viets 1938). The distribution of these eight genera is uncertain, they generally are excluded from further consideration.

Horizontal distribution

Geographical provinces with more than 100 halacarid species are the tropical western Pacific (TPW), temperate northeastern Atlantic (NAE), temperate

Table 1. Halacarid genera, number of known species (*n*) and number of species in geographical provinces.

Genus	*n*	ARC	NAE	NAW	MDB	TAE	TAW	SAE	SAW	NPE	NPW	TPE	TPW	SPE	SPW	TIE	TIW	SIE	ANT
Acanthohalacarus	1	–	1	–	–	–	–	–	–	–	–	–	–	–	–	–	–	–	–
Acanthopalpus	2	–	1	–	–	–	–	–	–	–	1	–	–	–	–	–	–	–	–
Acarochelopodia	9	–	1	–	4	1	–	–	–	–	–	3	2	–	–	–	1	–	–
Acaromantis	10	–	2	1	2	–	–	–	–	–	–	5	1	–	–	–	–	–	–
Acarothrix	3	–	–	–	–	–	–	–	–	–	–	–	2	–	–	–	1	–	–
Actacarus	24	–	3	1	3	–	3	–	–	1	4	4	5	3	–	–	1	2	–
Agaue	42	–	6	1	2	2	3	2	2	3	3	–	5	3	3	–	2	7	9
Agauides	2	–	1	–	–	–	–	–	–	–	–	–	1	–	–	–	–	–	–
Agauopsis	74	–	8	1	8	1	4	3	2	4	2	2	15	17	5	1	4	5	4
Anomalohalacarus	18	–	9	3	6	–	–	–	–	–	1	–	–	–	–	–	–	1	–
Arenihalacarus	1	–	–	–	–	–	–	–	–	–	1	–	–	–	–	–	–	–	–
Arhodeoporus	30	–	4	2	2	–	3	–	–	–	–	1	10	1	–	1	5	4	–
Atelopsalis	6	–	3	–	–	–	–	–	–	–	–	1	1	–	–	1	1	–	–
Australacarus	5	–	–	–	–	–	–	–	–	–	–	–	3	1	–	–	–	1	–
Bathyhalacarus	11	–	1	1	–	1	–	–	2	–	1	–	2	1	2	–	–	–	–
Bradyagaue	18	2	1	–	–	1	–	–	–	1	3	–	–	5	–	–	1	1	4
Camactognathus	3	–	1	–	1	–	–	–	–	–	–	–	–	1	–	–	–	–	–
Caspihalacarus	1	–	1	–	1	–	–	–	–	–	–	–	–	–	–	–	–	–	–
Coloboceras	3	–	3	–	1	–	–	–	–	–	–	–	–	–	–	–	–	–	–
Colobocerasides	2	–	1	–	–	–	–	–	–	–	–	–	–	–	–	–	–	–	1
Copidognathides	3	–	1	–	1	–	–	–	–	–	–	–	1	–	–	–	–	–	–
Copidognathus	336	13	38	9	33	3	32	4	6	24	33	18	57	45	4	12	23	27	13
Corallihalacarus	1	–	–	–	–	–	–	–	–	–	–	–	1	–	–	–	–	–	–
Enterohalacarus	1	–	–	–	–	–	–	–	–	–	–	–	1	–	–	–	–	–	–
Halacarellus	49	5	10	5	7	–	1	–	–	5	1	–	1	8	2	–	1	2	17
Halacaroides	2	–	–	–	–	–	–	–	–	–	–	–	–	2	–	–	–	–	–
Halacaropsis	5	–	1	–	1	–	–	–	–	–	–	–	1	–	1	–	1	1	–
Halacarus	74	–	7	3	6	2	3	–	8	2	2	–	11	7	4	–	5	8	16
Halixodes	3	–	–	–	–	–	–	–	–	–	–	–	–	–	3	–	–	–	–

Table 1. Continued

Genus	*n*	ARC	NAE	NAW	MDB	TAE	TAW	SAE	SAW	NPE	NPW	TPE	TPW	SPE	SPW	TIE	TIW	SIE	ANT
Isobactrus	26	1	8	4	4	–	–	–	–	–	6	2	3	2	–	–	–	–	3
Lohmannella	34	1	6	1	9	1	–	–	2	1	1	–	1	7	1	–	1	1	8
Metarhombognathus	2	2	2	2	–	–	–	–	–	–	–	–	–	–	–	–	–	–	–
Mictognathus	3	–	–	–	–	–	–	–	1	–	–	–	1	–	–	–	–	–	1
Parhalixodes	2	–	–	–	1	–	–	–	–	–	–	–	–	1	–	–	–	–	–
Pelacarus	1	–	1	–	1	–	–	–	–	–	–	–	–	–	–	–	–	–	–
Peregrinacarus	2	–	–	–	–	–	–	–	1	–	–	–	–	–	–	–	–	–	1
Phacacarus	1	–	–	–	–	–	–	–	–	–	–	–	–	–	–	–	–	1	–
Rhombognathides	6	2	6	4	1	–	–	–	–	–	–	–	–	–	–	–	–	–	–
Rhombognathus	93	1	6	2	10	–	1	1	1	2	12	8	27	8	3	2	1	10	8
Scaptognathides	10	–	–	–	–	–	3	–	–	–	–	3	4	–	–	–	–	1	–
Scaptognathus	28	–	4	–	4	–	1	–	–	–	3	3	9	3	1	1	4	2	–
Simognathus	38	–	2	1	2	–	1	1	1	–	–	2	13	6	2	–	3	7	–
Spongihalacarus	1	–	–	–	–	–	–	–	–	–	–	–	1	–	–	–	–	–	–
Thalassacarus	1	–	–	–	–	–	–	–	–	1	1	–	–	–	–	–	–	–	–
Thalassarachna	14	3	8	4	2	–	–	–	–	–	–	–	–	–	–	–	–	–	–
Thalassophthirius	1	–	–	–	–	–	–	–	1	–	–	–	–	–	–	–	–	–	–
Tropihalacarus	2	–	–	–	–	–	–	–	–	–	–	–	2	–	–	–	–	–	–
Werthella	10	–	–	–	–	1	–	–	1	–	–	–	–	3	1	–	–	1	4
Werthelloides	1	–	–	–	–	–	–	–	–	–	–	–	–	–	–	–	1	–	–
Winlundia	2	–	–	–	–	–	–	–	–	–	–	–	–	2	–	–	–	–	–
Xenohalacarus	1	–	–	–	–	–	–	–	–	–	–	–	1	–	–	–	–	–	–

Abbreviations cf. Figure 1; –, no record.

southeastern Pacific (SPE), and the Mediterranean-Black Sea-Aral Sea area (MDB) with 181, 147, 127 and 112 species, respectively, or 18, 14, 12 and 11% of all documented halacarid species. Noteworthy is the difference between the number of species between the northwestern and northeastern Atlantic, 147 versus 45 species, as well as that difference between the northwestern and northeastern Pacific, 75 versus 44 species. Provinces with scattered records are the temperate southeastern Atlantic (SAE), the tropical eastern Atlantic (TAE) and the tropical eastern Indian Ocean (TIE), with 11, 13 and 18 species, respectively; however, these are regions that have been relatively poorly sampled.

Several species are recorded from more than one of the geographical provinces. Examples of wide-spread species are: *Copidognathus curtus* Hall 1912, recorded from coastal Chile, Peru, USA and Japan (Newell 1984); *Thalassacarus commatops* Newell 1949 which covers a latitudinal area from California to the Aleutians (Newell 1949, 1951); *Actacarus pacificus* Bartsch 1979, obviously common in the Indo-Pacific region (Abé 1997; Bartsch 2003b) and *Rhombognathus auster* Bartsch 1989, spread in the southern Atlantic and Indian Ocean. From the Indian Ocean, Chatterjee (1995) and Sarma and Chatterjee (1991, 1993) presented records of species (*Arhodeoporus bonairensis* (Viets 1936), *Copidognathus hartwigi* Bartsch 1979, *C. longispinus* Bartsch and Iliffe, 1985) that were originally described from the Atlantic Ocean. These examples refer to upper littoral species. A deep-sea species with a large latitudinal range is *Agaue corollata* Bartsch 1978, with records from the Angola basin, 9° south, to the Labrador Basin, 58° north (Bartsch 1982b, 1999c).

When concentrating on the temperature influenced littoral provinces, tropical, temperate and polar, and excluding all records from a depth beyond 200 m (Table 2), then most species live in the tropics (325 species) and temperate northern oceans (309 species), followed by the temperate southern oceans (240 species). Significantly lower are the numbers of species recorded from polar and subpolar areas (30 species in the north, 73 in the south). On a generic level, 29 genera are known from tropical and temperate but not from polar provinces. Fourteen genera are exclusively from the temperate regions and five genera (three of them monotypic) are to date taken only from the tropics. None is restricted to polar regions.

The genera *Rhombognathus*, *Agauopsis*, *Actacarus* and *Scaptognathus* are similar abundant at temperate and tropical shores ($p > 0.05$). In *Arhodeoporus* and *Scaptognathides* more species live in the tropics than in the temperate zones; records from polar regions are lacking. Five genera (*Anomalohalacarus*, *Isobactrus*, *Metarhombognathus*, *Rhombognathides*, *Thalassarachna*) are more abundant in the north than in the south. One of these genera (*Metarhombognathus*) is restricted to the cold-temperate and polar northern Atlantic, and two genera (*Rhombognathides* and *Thalassarachna*) to the northern Atlantic and Mediterranean-Black Sea area. The record of *Rhombognathides pascens* (Lohmann 1889) from the Indian Ocean (TIW) (André 1959) is in need of verification and hence excluded from present considerations. In contrast, *Halacaroides*, *Peregrinacarus*, *Werthella*, and *Winlundia* are restricted to the

Table 2. Number of halacarid species recorded from different depth zones and number of shallow water (0–200 m) species in northern polar, northern temperate, tropical, southern temperate, and southern polar areas.

Genus	Depth (m)					Shallow water areas (0–200 m)				
	0–50	51–200	0–200	201–1000	> 1000	North polar	North temp	Tropics	South temp	South polar
Acanthohalacarus	–	–	–	1	–	–	–	–	1	–
Acanthopalpus	2	–	2	–	–	–	2	–	–	–
Acarochelopodia	9	–	9	–	–	–	4	6	–	–
Acaromantis	10	–	10	–	–	–	4	6	–	–
Acarothrix	3	–	3	–	–	–	–	3	–	–
Actacarus	24	1	24	1	–	–	8	13	5	–
Agaue	33	8	35	9	4	–	8	10	14	6
Agauides	–	–	–	1	1	–	–	–	–	–
Agauopsis	66	7	67	7	1	–	16	25	27	4
Anomalohalacarus	18	4	18	–	–	–	17	–	1	–
Arenihalacarus	1	–	1	–	–	–	1	–	–	–
Arhodeoporus	26	4	28	4	1	–	5	18	5	–
Atelopsalis	2	1	3	4	1	–	1	2	–	–
Australacarus	3	2	5	–	–	–	–	3	2	–
Bathyhalacarus	–	1	1	1	9	–	–	–	1	–
Bradyagaue	8	8	13	9	3	2	3	1	5	2
Camactognathus	3	–	3	–	–	–	2	–	1	–
Caspihalacarus	1	–	1	–	–	–	1	–	–	–
Coloboceras	2	–	2	1	–	–	2	–	–	–
Colobocerasides	–	–	–	1	1	–	–	–	–	–
Copidognathides	3	–	3	–	–	–	2	–	–	–
Copidognathus	303	34	317	20	12	13	107	129	76	13
Corallihalacarus	1	–	1	–	–	–	–	1	–	–
Enterohalacarus	–	–	–	1	–	–	–	–	–	–
Halacarellus	35	10	43	7	2	5	18	3	10	14
Halacaroides	2	–	2	–	–	–	–	–	2	–
Halacaropsis	5	–	5	–	–	–	1	2	2	–
Halacarus	45	15	53	26	3	–	14	14	18	10
Halixodes	3	–	3	–	–	–	–	–	3	–
Isobactrus	26	–	26	–	–	1	17	5	2	3
Lohmannella	30	6	33	4	2	1	14	2	11	8
Metarhombognathus	2	–	2	–	–	2	2	–	–	–
Mictognathus	2	2	3	–	–	–	–	1	1	1
Parhalixodes	1	–	1	1	–	–	1	–	–	–
Pelacarus	1	1	1	1	1	–	1	–	–	–
Peregrinacarus	2	–	2	–	–	–	–	–	1	1
Phacacarus	1	–	1	–	–	–	–	–	1	–
Rhombognathides	6	1	6	–	–	2	6	–	–	–
Rhombognathus	91	2	93	–	–	1	31	36	23	8
Scaptognathides	10	–	10	–	–	–	–	9	1	–
Scaptognathus	25	2	25	4	–	–	8	13	5	
Simognathus	38	2	38	1	–	–	4	19	15	–
Spongihalacarus	1	–	1	–	–	–	–	1	–	–
Thalassacarus	1	–	1	–	–	–	1	–	–	–

Table 2. Continued

Genus	Depth (m)					Shallow water areas (0–200 m)				
	0–50	51–200	0–200	201–1000	> 1000	North polar	North temp	Tropics	South temp	South polar
Thalassarachna	11	3	11	1	2	3	8	–	–	–
Thalassophthirius	1	–	1	–	–	–	–	–	1	–
Tropihalacarus	2	–	2	–	–	–	–	2	–	–
Werthella	4	4	8	1	2	–	–	–	4	4
Werthelloides	–	–	–	–	1	–	–	–	–	–
Winlundia	2	–	2	–	–	–	–	–	2	–
Xenohalacarus	1	–	1	–	–	–	–	1	–	–

–, No record.

southern polar and/or southern temperate waters. In the genus *Simognathus* more species are recorded from the south and tropics than from the north, and *Agaue* and *Halacarus* are present in Antarctic and sub-Antarctic, southern temperate, tropical and northern temperate provinces, but there are no records from northern polar waters.

Thirteen genera are restricted to one of the biogeographical provinces, but 10 of these genera (*Arenihalacarus, Corallihalacarus, Halacaroides, Phacacarus, Spongihalacarus, Thalassophthirius, Tropihalacarus, Werthelloides, Winlundia, Xenohalacarus*) are sparsely recorded. The genus *Halixodes*, reported from New Zealand (SPW), is characterized by enlarged barbs on the chelicerae. Similar barbs have been found in a nymph from outside the given area (Bartsch 1986a).

The data presented above reflect the distribution of genera without consideration of phylogenetic lineages or natural groups within a genus. Several such species groups show distinct latitudinal or longitudinal distributional patterns. An example is the genus *Isobactrus*, spread world-wide but the northern species are distinct from the southern (Bartsch 2000; Abé 2001). Northern hemisphere *Lohmannella* differ from most southern species in the size of gland pores. *Anomalohalacarus* species from the Pacific and Indian Ocean are distinct from those from the northern Atlantic (Bartsch 2003b). Some of the species groups in the genera *Agauopsis, Arhodeoporus, Copidognathus*, and *Halacarellus* seem to be either bound to or excluded from a temperature zone. The *ornata* group in *Agauopsis*, the *bairdi* and *curassiviensis* group in *Copidognathus*, and the *bonairensis* and *eclogarius* group in *Arhodeoporus* are present only in warm-temperate and tropical waters (Bartsch 1983, 1984a, 1993, 1999a; Otto 2000c). Other species groups dominate or are restricted to one of the oceans.

Vertical distribution

As shown in Table 2, 866 species or 85% of halacarid mites are recorded from the upper 50 m. The number also includes the species extracted from algae washed ashore. When considering the shelf area to a depth of 200 m, then 920 or 90% of the halacarid species live here. No more than 46 species are recorded from the deep sea.

Eleven out of the 51 halacarid genera inhabit the range from shallow (0–50 m) to the deep sea (> 1000 m). After excluding the poorly collected monotypic genera 18 genera have been exclusively recorded from the upper 200 m (Table 2) and three genera (*Agauides*, *Bathyhalacarus*, *Colobocerasides*) have no or almost no records from the upper 200 m.

Salinity

Halacarid mites are basically marine but as shown in the Table 3, there are species adapted to life in fresh or oligohaline brackish water, at salinity of about 1.5‰ or less. The number given in the table includes only the species regularly or exclusively living in fresh water, sporadic records (xenobionts) from fresh water are omitted. Species penetrating into fresh water are found in the genera *Acarothrix*, *Caspihalacarus*, *Copidognathus*, *Halacarellus*, *Isobactrus*, *Lohmannella*, and *Peregrinacarus*.

Habitat

Information on habitat is often fragmentary and details on associations with macrofauna in general lacking. Eight genera are extracted from at least four of the five habitats mentioned. Exclusively arenicolous are *Acarochelopodia*, *Actacarus*, *Anomalohalacarus* and *Scaptognathides*, though they sporadically may be recorded from other habitats, e.g., *Anomalohalacarus* from the surface of echinoids (Bartsch 1989b). None of the halacarid genera is restricted to algal turf, fronds or hapteres. The three genera *Enterohalacarus*, *Parhalixodes* and *Spongihalacarus* are thought to be obligate associates of echinoids, nemerteans and sponges, respectively (Viets 1938; Laubier 1960; Otto 2000a). Two of these genera are monotypic.

Discussion

Before analyzing the results one should keep in mind that: (1) The tables are compiled from data reported in literature which are based on very different sampling activities. From many areas almost no records are available, from other areas, descriptions of the halacarid fauna has concentrated on a few genera. (2) The boundaries of the geographical provinces are artificial and not as strict as illustrated in the map (Figure 1). Surface currents and temperature are strongly underlying seasonal, annual or irregular changes, and taxa may react differently on fluctuations. Similarly, there are no abrupt boundaries in

Table 3. Number of halacarid species recorded from marine and diluted brackish water (≤ 1.5‰) and records (+) or no records (–) of a genus in the habitats: A = median-sorted sandy deposits rich in interstitia, B = endo- and epifauna amongst debris in sediments and burrows of a macrofauna, C = algal turf and fronds, D = colonial organisms (e.g. bryozoans, hydrozoans, barnacles, serpulids), E = association with a macrofauna (e.g., sponges, decapods, echinoderms.

Genus	Salinity		Habitat				
	Marine	Brackish (≤ 1.5‰)	A	B	C	D	E
Acanthohalacarus	1	–	–	+?	–	–	?
Acanthopalpus	2	–	+	+	–	–	–
Acarochelopodia	9	–	+	–	–	–	–
Acaromantis	10	–	+	+	–	–	–
Acarothrix	3	1(–3)	–	+	–	–	–
Actacarus	24	–	+	–	–	–	–
Agaue	42	–	–	+	+	+	–
Agauides	2	–	–	+?	–	–	?
Agauopsis	74	–	+	+	+	+	–
Anomalohalacarus	18	–	+	–	–	–	(+)
Arenihalacarus	1	–	+	–	–	–	–
Arhodeoporus	30	–	+	+	+	+	(+)
Atelopsalis	6	–	+	+	–	–	–
Australacarus	5	–	–	+	–	–	?
Bathyhalacarus	11	–	–	+?	–	+	–
Bradyagaue	18	–	–	–	–	+	–
Camactognathus	3	–	–	+	?	–	–
Caspihalacarus	1	1	–	+	–	–	–
Coloboceras	3	–	+?	+?	–	–	+?
Colobocerasides	2	–	–	–	–	+?	+?
Copidognathides	3	–	+	+	+	–	–
Copidognathus	232	5	+	+	+	+	+
Corallihalacarus	1	–	–	+	–	–	–
Enterohalacarus	1	–	–	–	–	–	+
Halacarellus	48	6	+	+	+	+	+
Halacaroides	2	–	+?	+?	–	–	–
Halacaropsis	5	–	–	–	+	+	–
Halacarus	74	–	+	+	+	+	+
Halixodes	3	–	–	–	–	+	+
Isobactrus	26	2	–	–	+	+	–
Lohmannella	29	5	+	+	+	+	–
Metarhombognathus	2	–	–	–	+	+	–
Mictognathus	3	–	+	+	–	–	–
Parhalixodes	2	–	–	–	–	–	+
Pelacarus	1	–	–	+	–	–	–
Peregrinacarus	–	2	–	+	–	–	–
Phacacarus	1	–	–	–	+	–	–
Rhombognathides	6	–	+	+	+	–	–
Rhombognathus	93	–	+	+	+	–	–
Scaptognathides	10	–	+	–	–	–	–
Scaptognathus	28	–	+	+	–	–	–
Simognathus	38	–	+	+	+	–	–
Spongihalacarus	1	–	–	–	–	–	+
Thalassacarus	1	–	–	–	+	–	–

Table 3. Continued

Genus	Salinity		Habitat				
	Marine	Brackish (≤ 1.5‰)	A	B	C	D	E
Thalassarachna	14	–	–	+	+	+	+
Thalassophthirius	1	–	–	?	?	?	?
Tropihalacarus	2	–	+	+	–	–	–
Werthella	10	–	+	+	+	+	–
Werthelloides	1	–	–	+	–	–	–
Winlundia	2	–	+	+	+	–	–
Xenohalacarus	1	–	+	–	–	–	–

Rare records are in parentheses; question mark indicates a possible but unproven association.

respect to the depth zones. (3) The taxonomic classification is man-made. Though most of present genera seem to be reliably defined, the position of perhaps a dozen species is questioned and even the rank of the one or more genera may have to be discussed.

Number of genera and species

It is noteworthy that a single genus, *Copidognathus*, holds almost one-third of the halacarid fauna. The genus is defined with help of a strict combination of characters (cf. diagnosis in Bartsch 1994) according to a Darwinian classification. There are no reports on species with doubtful position within this genus. In recent years several natural species groups have been described (Bartsch 1984a, 1994, 1997a, c, 1999d; Otto 2001), but these groups have never been treated as subgenera and their typical combination of characters do not require or support a subdivision of the genus *Copidognathus*. In the large aquatic mite group in fresh water, the Hydrachnidia, there is no such dominance of a single genus, and genera rich in species are divided into subgenera (Viets 1987).

Horizontal distribution

The data reveal a distinct predominance of the temperate and tropical halacarid fauna over that of polar waters. Certainly, fieldwork on the shores and shallows of polar oceans and adjacent basins and rivers will add numerous species, but increased sampling in temperate and tropical regions is likely to result in a much larger number of records. It seems unlikely that polar and subpolar waters will reach the number of species achieved in lower latitudes. Much of the data accumulated in Tables 1 and 2 are from tidal and shallow subtidal zones; in Arctic and Antarctic provinces the fauna in these zones will suffer from unpredictable disturbance and probably also a prolonged generation time, as documented in the littoral mite *Ameronothrus lineatus* (Thorell 1871) (Oribatida) (Søvik 2002). Data from greater depth and comparable substrata but different latitudes are lacking.

Is there a latitudinal diversity cline, with an increase of species richness with decreasing latitude, as postulated by Brown (1988) and Rosenzweig (1997)? Comparison of data (Table 2) from temperate and tropical waters demonstrate, more halacarid genera are recorded from the temperate provinces than from the tropics, fourteen genera are exclusively temperate, five genera tropical; three of the latter are monotypic, collected just once, and hence their life style is hardly known. On species level, there is no marked difference in number of halacarids between northern temperate and tropical waters, but slight difference between tropical and southern temperate. The author had the opportunity to study a tropical and warm-temperate area of Western Australia. The number of species in the warm-temperate Rottnest Island, off Perth (78 species), is more than twice as high as that of the tropical Dampier (unpublished), as well as that of the tropical Hong Kong (35 species, Huang 1994; Bartsch 1997a). High numbers of species have also been recorded from cold-temperate provinces. An example, the fauna of Roscoff, French coast of the English Channel, contains at least 70 halacarid species (Monniot 1964, 1967; Bartsch 1979, 1980, 1984b, 1991).

Like many gammaridean amphipods (Barnard 1991; Barnard and Karaman 1991), halacarids do not follow the general trend of a latitudinal biodiversity cline. But, in contrast to the amphipods, halacarid diversity in polar regions is low. Instead of a latitudinal cline from tropical to temperate zones, halacarids demonstrate a small scale species richness ('hot spots'), as also shallow water amphipods and soft sediment faunas (Myers 1996; Gray 2002). The diversity of marine mites will highly depend on antiquity of the region, environmental parameters, substrate heterogeneity and frequency of unpredictable disturbances.

A comparison of cold-temperate faunas of either side of the Atlantic showed that there are more than twice the number of species in the eastern than in the western Atlantic, despite their sharing a similar temperature regime. In 1982, the number of shallow water species amounted to 91 in the boreal Europe versus 38 in North America (Bartsch 1982a). During studies in the recent two decades, more species have been added to the fauna of the coldwater northeastern than to the northwestern Atlantic. Why is there such a significant difference in the number of species? An explanation for the markedly lower number in the east may be that parts of the Tethyan fauna never reached the opposite coast of the evolving Atlantic Ocean. Another reason can be found in the last glaciation, that after the retreat of the ice and rising sea level, shores of the eastern Atlantic were more quickly recolonized from the south due to adequate water currents. Noteworthy is the high number of species in common; 45% of the cold-temperate western Atlantic species also live in the eastern Atlantic (Bartsch 1982a). Shared species are restricted to the genera *Rhombognathides*, *Metarhombognathus*, *Isobactrus*, *Halacarellus*, and *Thalassarachna*. The species composition of *Copidognathus*, *Halacarus*, *Anomalohalacarus*, *Arhodeoporus*, *Agauopsis*, and *Rhombognathus* is different on either side. The amphiatlantics are thought to be ancient species with a high ability to adapt to changes in the environment. Resistant to harsh arctic conditions, they may have survived the ice covering of the last glaciation and quickly recolonized the emergent shorelines after retreat of the ice, both in the eastern and western Atlantic. Other species, today being a part of the cold-temperate fauna, may have come from southern populations which in the millions of years the shorelines were separated had evolved and speciated differently on either side of the northern Atlantic Ocean.

From the northern Pacific fauna no similar data are available. Studies on the shallow water halacarid fauna of either side often concentrated on different genera.

Vertical distribution

When looking at Table 2, one may get the impression that the majority of the genera are restricted to a rather narrow depth range. Can that be true?

The phytophagous genera *Isobactrus*, *Metarhombognathus*, *Rhombognathides*, and *Rhombognathus* have a depth distribution which is bound to the presence of algae, and the deepest record of crustose algae is from 268 m (Littler et al. 1985). *Isobactrus* and *Metarhombognathus* seem to be restricted to a horizon regularly or irregularly emerged, whereas *Rhombognathides* and *Rhombognathus* inhabit regions from the upper littoral to about 50 m. The deepest record is from 148 m (*Rhombognathus sandwichi* Newell 1984). The brackish water genera *Acarothrix*, *Caspihalacarus* and *Peregrinacarus* are expected to be found mainly in areas with fluctuating salinity, and hence to be bound to the upper 50 m. Several of the arenicolous genera are predominantly or exclusively found in an upper zone of 0–50 m. But as suggested by the records of *Actacarus* from about 400 m, and *Anomalohalacarus*, *Scaptognathides* and *Simognathus* from about 500 m (Bartsch 1977, 2003a), any absence of records from the continental slope and deep sea just reflects the scarce number of samples from adequate sediment and depth.

Consequently, apart from genera bound to special food items or habitats (algae or brackish water), the littoral halacarid genera are expected to penetrate beyond the shelf edge. Similarly, genera from great depth (*Bathyhalacarus*) have species on the shelf.

Is the fauna more diverse in the upper 50 or 200 m than on the slope and in the deep sea? The species recorded from the tidal and shelf area by far outnumber those found at greater depth (Table 2), but so does the number of collections. Newell (1971) mentioned 14 species (93 individuals) at a depth of 62–160 m, 18 species (443 individuals) at 485–660 m, and seven species below 1565 m (33 individuals). He calculated the number of species in the temperate southeastern Pacific to amount to about 40 in the tidal but to 54 in the zone beyond 60 m. Newell (1971) expected a higher number of halacarid species in the deep than in the shallow zones to be true for the world as a whole.

The calculations by Newell (1971) partly agree with the depth gradient generally thought to exist in marine organisms, with an increase of number of species with depth, and both meio- and macrofauna taxa having the highest species diversity below 1000 m (Coull 1972; Rex 1981; Brown 1988; Zezina 1997), and often having a maximum at depths of 2000–3000 m. The present data on halacarid fauna do not follow the just mentioned trend. The contrast between halacarids and most other marine taxa may be due to differences in the phylogenetic history. Mites are primarily terrestrial and the halacarids are expected to have evolved from semi-aquatic marine ancestors which once colonized the sea shores (Bartsch 1996a), and since diversified and penetrated towards the hadal. Families closely related to the Halacaridae are the prostigmatids Cunaxidae and Bdellidae, which are terrestrial though many of them are found close to a water line; the prostigmatids Pontarachnidae are present only in the shallow subtidal zone. Representatives of the suborders Oribatida (e.g. Ameronothroidea), Astigmata (Hyadesiidae), and Mesostigmata (Rhodacaridae, Halarachnidae) live in the tidal zone or are parasites in the nasal passages of seal and walrus. Amongst the suborders and families of Acari, the family Halacaridae is the only one which invaded all depths of the oceans.

The dominance of halacarid species recorded from the littoral zone does not imply that species diversity is low in the bathyal. The fauna can be as rich as that of many shallow water areas. Recent meiofauna collections of the Great Meteor Seamount (northeastern Atlantic) raised the number of species of this isolated seamount beyond 25 (Bartsch, unpublished data). That means, the bathyal fauna may be as diverse as that of many tropical and temperate littoral areas, though, at a global scale, it may not reach the number of species recorded from littoral 'hot spots'.

Whereas most genera demonstrate a wide depth distribution, species often do not. Some few species have been recorded from a range from tidal to beyond 1000 m, viz. *Agaue parva* (Chilton 1883), *Arhodeoporus gracilipes* (Trouessart 1889), *Bradyagaue drygalskii* (Lohmann 1907), *Copidognathus oculatus* (Hodge 1863), *C. posticus* Newell 1971, and *Lohmannella falcata* (Hodge 1863). Are the data reliable? Records of *A. gracilipes*, *B. drygalskii*, *C. oculatus*, and *L. falcata* have often been based on mis-identified specimens (Newell 1984; Bartsch 2003a). *Agaue parva* from the tidal zone (Chilton 1883) is not identical with the species described by Newell (1984) and currently accepted to represent *A. parva* (the specimen said to be the holotype of *A parva* proved not to resemble *A. parva* sensu Newell, Bartsch, pers. observ.). The shallow water record of *C. posticus* (Newell 1984: 197) is most likely a (printing) lapse. Halacarid species may inhabit a range from the sublittoral to bathyal, but populations of tidal species are not expected to penetrate beyond 1000 m.

Endemisms

Tables 1 and 2 show that most genera are recorded from more than one province and often more than one depth zone. Eleven out of the thirteen genera restricted to one of the biogeographical provinces are represented by fewer than 20 specimens, and the absence of records of these genera from anywhere else is more likely a question of paucity of sampling than an evidence for endemism. Of the remaining two genera, *Acanthohalacarus* and *Enterohalacarus*, the first was recently taken on the Great Meteor Seamount (Bartsch 2001a) and similar samples from the continental slope of Africa are expected to bring more specimens. The other genus is known to be bound to echinoids and detailed analyses of echinoids from the sublittoral to the abyssal are expected to add new records.

Within the taxon Halacaridae, endemisms on a genus level seem to be rare or even absent. This assumption is corroborated by the distribution of genera such as *Agauides*, with no more than two specimens recorded, one from the Atlantic, one from the Pacific (Bartsch 1988a, 1989c), *Parhalixodes*, with two individuals recorded from the Mediterranean and one from the Pacific (Laubier 1960; Newell 1971), and *Peregrinacarus* and *Acarothrix*, brackish/fresh water genera, the former known from the Marion and Falkland Islands (Bartsch 1999b, 2001b), the latter from the South China Sea (Hong Kong, southern China), Timor Sea (Darwin, Australia), southwestern Indian Ocean (Durban, South Africa) and Gulf of Mexico (Tampa, Florida) (Bartsch 1990, 1997b, and unpublished records; Proches 2002). On a species level, small scale endemisms will occur, as demonstrated by *Rhombognathus cetratus* Bartsch 1974 (Bartsch 1996a).

Means of dispersal

Halacarid mites, small and without dispersal stages, are dependant on passive transport. Rafting is a common way for spreading, as documented for macrofaunal species (Hobday 2000). Though halacarids are likely to be washed out from drifting algae rather quickly, dispersal via rafts should still be possible. Shallow water inhabitants can be transported by storms (cf. Baijkov 1949), water fowl or insects. The two latter means are known or expected to carry with them both aquatic macro- and meiofauna (Maguire 1963; Bartsch 1995, 1996a; Milbrink 1999; Wilhelmsen 1999; Bilton et al. 2001). In the bathyal and deep-sea, halacarid mites can be transported with sediment removed by underwater 'storms' and currents and by debris trapped on the tegument of errant macrofauna. In the recent centuries, with increased world-wide travel and transport of goods, thousands of species have been transported to foreign coasts, halacarids as well (see below).

Introduced species

The small halacarid mites generally escape notice and hence the knowledge of the halacarid fauna of major coastlines is fragmentary. Records of species far from previously known localities may raise the question whether the new record lies within the native range of a given species or if the species has been introduced.

Isobactrus uniscutatus (Viets 1939), recorded from southeastern Australia, is a species transported to Australia, as inferred from phylogenetic, biological and ecological data (Bartsch and Gwyther 2004).

Caspihalacarus hyrcanus Viets 1928, formerly thought to be bound to the Ponto-Caspian area, was extracted from the Rhine in the Netherlands (Bij de Vaate et al. 2002). The aquatic fauna of the Netherlands has been studied intensely in the past two centuries and it is unlikely that this rather large-sized species (length 450–540 μm) had escaped notice. The colonization of the Rhine likely occurred recently.

In the past decades many marine species have been introduced into the Baltic (Leppäkoski et al. 2002). In respect to the taxon Halacaridae, *Copidognathus brachystomus* Viets 1940 may be such a species new to the Baltic fauna. The species is abundant in the Mediterranean and Black Sea (Morselli 1980; Bartsch 2001c), it has been taken by the author on the coast of Mecklenburg (eastern Germany), on the islands Hiddensee (in 1989) and Poel (in 2002). In the latter area the species was present in large numbers amongst the ooze on and between shallow water green algae (*Cladophora*). Studies on the Baltic halacarid fauna began in the 19th century (Lohmann 1889) and went on in the past century, but there had been no records of *C. brachystomus*.

Isobactrus uniscutatus, *I. ungulatus* Bartsch 1975, and *Thalassarachna basteri* (Johnston 1836) may be alien species in the Mediterranean fauna. The two former species have been collected in the Adriatic Sea, the latter in the Tyrrhenian Sea (Viets 1939; Bartsch 1976, 1998).

Conclusion

Halacarid mites are spread world wide. Based on current records, the highest diversity occurs in the tropical and temperate provinces, whereas the number of species is small in polar regions. Halacarid distribution shows no latitudinal gradient, with a decrease in species richness with increasing latitude, as is expected to be an almost universal trend for the majority of taxa. The largest number of halacarid species have been found in the provinces tropical western Pacific (TPW), temperate northeastern Atlantic (NAE), temperate southeastern Pacific (SPE), and Mediterranean-Black Sea (MDB). At 'hot spots', the number of species from cold or warm-temperate areas may more than double that from the tropics. Species richness of halacarids seems to be dependant on regional environmental parameters, substrate heterogeneity and antiquity.

According to present data, the halacarid fauna is distinctly more diverse in the upper 200 m than in the bathyal. The number of species from a depth exceeding 1000 m is negligible. These findings are in contrast to the trend that greatest species diversity occurs not in the shallowest water but at a depth of about 2000–3000 m.

Only a few genera appear to be bound to littoral habitats; these are genera that include algivorous and brackish/fresh water species. Most genera are expected to have representatives in littoral, bathyal and even abyssal depths.

There is little evidence that any of the halacarid genera are endemic to one of the marine geographical provinces. Halacarids are thought to be an ancient group with several genera dating back to the Mesozoic or even Pre-Mesozoic and, inferred from data on geographical distribution and biology, species of some of the genera are expected to have survived unchanged for more than 50 million years. Even if the chance of dispersal and the fecundity of halacarids is low, they had millions of years to be spread on the globe and in that length of time, they could afford numerous failures in colonization.

References

Abé H. 1991. A new genus and species of the family Halacaridae (Acari, Prostigmata) from Japan. Zool. Jb. (Syst.) 118: 247–256.

Abé H. 1997. Halacarid mites of the genus *Actacarus* (Acari: Halacaridae) from Hokkaido, Northern Japan. Sp. Div. 2: 31–42.

Abé H. 2001. Phylogeny and character evolution of the marine mite genus *Isobactrus* (Acari : Halacaridae). J. Nat. Hist. 35: 617–625.

André M. 1959. XXVI. - Acari I. Contribution a l'étude des halacariens de la Mer Rouge. Mission Robert Ph. Dollfus en Égypte, 93–119.

Baijkov A.D. 1949. Do fish fall from the sky? Science 109: 402.

Barnard J.L. 1991. Amphipodological agreement with Platnick. J. Nat. Hist. 25: 1675–1676.

Barnard J.L. and Karaman G.S. 1991. The families and genera of marine gammaridean Amphipoda (except marine gammaroids) Part 1. Rec. Austr. Mus. Suppl. 13: 1–417.

Bartsch I. 1972. Ein Beitrag zur Systematik, Biologie und Ökologie der Halacariden (Acari) aus dem Litoral der Nord- und Ostsee. I. Systematik und Biologie. Abh. Verh. naturwiss. Ver. Hamburg (NF) 16: 155–230.

Bartsch I. 1976. Ergnzungen zur Halacariden-Fauna (Halacaridae, Acari) im Becken von Arcachon. Vie Milieu, A 26: 31–46.

Bartsch I. 1977. Eine neue *Actacarus*-Art (Acari, Halacaridae) aus dem Bathyal vor der Küste von North Carolina, USA. Zool. Scr. 6: 323–326.

Bartsch I. 1979. Verbreitung der Halacaridae (Acari) im Gezeitenbereich der Bretagne-Küste, eine ökologische Analyse. II. Quantitative Untersuchungen und Faunenanalyse. Cah. Biol. Mar. 20: 1–28.

Bartsch I. 1980. Halacaridae (Acari) aus der Bucht von Morlaix (Bretagne). Acarologia 21: 34–45.

Bartsch I. 1982a. Halacaridae (Acari) von der Atlantikküste des borealen Nordamerikas. Ökologische und tiergeographische Faunenanalyse. Helgoländer Meeresunters. 35: 13–46.

Bartsch I. 1982b. Zur *corrolata*-Gruppe der Gattung *Agaue*, Verbreitung und Beschreibung von zwei neuen Arten. Acarologia 23: 45–53.

Bartsch I. 1983. Zur Systematik und Verbreitung der Gattung *Arhodeoporus* (Halacaridae, Acari) und Beschreibung zweier neuer Arten. Zool. Beitr. N.F. 28: 1–16.

Bartsch I. 1984a. New species of the *bairdi* group in the genus *Copidognathus* (Acari, Halacaridae). Bull. Mar. Sci. 35: 200–210.

Bartsch I. 1984b. Ergänzungen zur Halacariden-Fauna der Bretagne-Küste und Beschreibung einer neuen Art (Halacaridae, Acari). Cah. Biol. Mar. 25: 113–122.

Bartsch I. 1986a. A new species of *Halixodes* (Halacaridae, Acari) and a review of the New Zealand species. J. Roy. Soc. N. Z. 16: 51–56.

Bartsch I. 1986b. Zur Gattung *Werthella* Lohmann, *Pelacarus* n. gen. und *Werthelloides* n. gen. (Halacaridae, Acari). Cah. Biol. Mar. 27: 211–223.

Bartsch I. 1988a. Halacariden (Acari) im Nordatlantik. Beschreibung von *Halacarus balgimus* sp. n. und *Agauides cryosi* gen. et sp. n. Cah. Biol. Mar. 29: 353–359.

Bartsch I. 1988b. *Thalassophthirius auster* gen. et spec. nov., a halacarid mite (Acari) suspected of being a parasite. Polar Res. 6: 181–184.

Bartsch I. 1989a. Süsswasserbewohnende Halacariden und ihre Einordnung in das System der Halacaroidea (Acari). Acarologia 30: 217–239.

Bartsch I. 1989b. *Copidognathus brifacius* n. sp. (Halacaridae, Acari) und Bemerkungen zu weiteren Bewohnern des Seeigels *Spatangus purpureus* (Echinoidea). Mitt. Hamb. Zool. Mus. Inst. 86: 315–328.

Bartsch I. 1989c. Deep-sea mites (Halacaridae, Acari), from the southwestern Pacific. Cah. Biol. Mar. 30: 455–471.

Bartsch I. 1990. *Acarothrix palustris* gen. et. spec. nov. (Halacaroidea, Acari), ein Bewohner der Salzwiesen Südchinas. Zool. Anz. 224: 204–210.

Bartsch I. 1991. On the identity of some North Atlantic halacarid species. J. Nat. Hist. 25: 1339–1353.

Bartsch I. 1992a. *Phacacarus flavellus* gen. et spec. nov. (Copidognathinae, Halacaroidea, Acari), eine Meeresmilbe auf Kalkalgen. Zool. Anz. 228: 212–219.

Bartsch I. 1992b. Hong Kong rhombognathine mites (Acari: Halacaridae). In: Morton B. (ed.), The Marine Flora and Fauna of Hong Kong and Southern China, III. University Press, Hong Kong, pp. 251–276.

Bartsch I. 1993. Halacaridae (Acari) von Malaysia. Beschreibung von drei Arten der Gattung *Copidognathus*. Ent. Mitt. Zool. Mus. Hamburg 11: 45–58.

Bartsch I. 1994. *Copidognathus* (Halacaridae: Acari) from Western Australia. Description of twelve new species of the *gibbus* group. Rec. West. Aust. Mus. 16: 535–566.

Bartsch I. 1995. A new subspecies of the freshwater halacarid mite *Lobohalacarus weberi* (Romijn and Viets) (Halacaridae, Acari) from a Southern Atlantic Ocean island. Ann. Cape Prov. Mus. (nat. Hist) 19: 171–180.

Bartsch I. 1996a. Halacarids (Halacaroidea, Acari) in freshwater. Multiple invasions from the Paleozoic onwards. J. Nat. Hist. 30: 67–99.

Bartsch I. 1996b. Halacarines (Acari: Halacaridae) from Rottnest Island, Western Australia: the genera *Agauopsis* Viets and *Halacaropsis* gen. nov. Rec. West. Aust. Mus. 18: 1–18.

Bartsch I. 1997a. New species of the *Copidognathus gibbus* group (Acari: Halacaridae) from Hong Kong. In: Morton B. (ed.), The Marine Flora and Fauna of Hong Kong and Southern China IV. University Press, Hong Kong, pp. 63–76.

Bartsch I. 1997b. Copidognathinae (Halacaridae, Acari) from northern Australia; description of four new species. In: Hanley J.R., Caswell G., Megirian D. and Larson H.K. (eds), Proceedings of the Sixth International Biological Workshop. The marine flora and fauna of Darwin Harbour. Northern Territory, Australia, Western Australian Museum, Perth, pp. 231–243.

Bartsch I. 1997c. A new species of the *Copidognathus tricorneatus* group (Acari: Halacaridae) from Western Australia with a review of this species-group. Sp. Div. 2: 155–166.

Bartsch I. 1998. Halacarinae (Acari: Halacaroidea) from the northwestern Black Sea: a review. Mitt. Hamb. Zool. Mus. Inst. 95: 143–178.

Bartsch I. 1999a. Wiederbeschreibung zweier Arten der *Agauopsis ornata*-Gruppe (Acari, Halacaridae). Ent. Mitt. Zool. Mus. Hamburg 13: 37–48.

Bartsch I. 1999b. *Peregrinacarus reticulatus* gen. et. spec. nov., a freshwater halacarid mite (Acari, Halacaridae) from Marion Island. Hydrobiologia 392: 225–232.

Bartsch I. 1999c. Halacaridae (Acari) from Rottnest Island. Description of two *Agaue* species. Acarologia 40: 179–190.

Bartsch I. 1999d. *Copidognathus* (Halacaridae: Acari) from western Australia: five species of the *oculatus* group. Rec. West. Aust. Mus. 19: 299–321.

Bartsch I. 2000. Rhombognathinae (Acari: Halacaridae) from the Great Barrier Reef, Australia. Mem. Qd Mus. 45: 165–203.

Bartsch I. 2001a. A new halacarid genus (Acari: Halacaridae: Halacarinae) from the Great Meteor Seamount, Eastern North Atlantic. Sp. Div. 6: 117–125.

Bartsch I. 2001b. A new freshwater halacarid mite, genus *Peregrinacarus* (Halacaridae, Acari) from the Falklands. Hydrobiologia 452: 139–144.

Bartsch I. 2001c. Black Sea Copidognathinae (Arachnida: Acari: Halacaridae): a review. Mitt. Mus. Nat. kd Berl., Zool. Reihe 77: 247–275.

Bartsch I. 2003a. Lohmannellinae (Halacaridae: Acari) from the Great Meteor Seamount (Northeastern Atlantic): description of new species and reflections on the origin of the seamount fauna. Mitt. Hamb. Zool. Mus. Inst. 100: 101–117.

Bartsch I. 2003b. Psammophilous halacarids (Halacaridae, Acari) from Dampier, Western Australia. Description of species and faunal comparison of the mesopsammal halacarid fauna of western and eastern Australia. West. Aust. Mus. 22: 23–45.

Bartsch I. and Gwyther J. 2004. A non-indigenous halacarid species in Victoria, southeastern Australia, *Isobactrus uniscutatus* (Rhombognathinae, Halacaridae, Acari). Proc. Soc. Victoria 116: 19–23.

Bij de Vaate A., Jazdzewski K., Ketelaars H.A.M., Gollasch S. and Van der Velde G. 2002. Geographical patterns in range extension of Ponto-Caspian macroinvertebrate species in Europe. Can. J. Fish. Aquat. Sci. 59: 1159–1174.

Bilton D.T., Freeland J.R. and Okamura B. 2001. Dispersal in freshwater invertebrates. Annu. Rev. Ecol. Syst. 32: 159–181.

Bolton J.J., Leliaert F, De Clerck O., Anderson R.J., Stegenga H., Engledow H.E. and Coppejans E. 2004. Where is the western limit of the tropical Indian Ocean seaweed flora? An analysis of intertidal seaweed biogeography on the coast of South Africa Mar. Biol. 144: 51–59.

Briggs J.C. 1995. Global Biogeography. Elsevier, Amsterdam.

Brown J.H. 1988. Species diversity. In: Myers A.A. and Giller P.S. (eds), Analytical Biogeography. An Integrated Approach to the Study of Animal and Plant Distribution. Chapman and Hall, London, pp. 57–89.

Chatterjee T. 1995. Occurence of *Copidognathus longispinus* Bartsch and Iliffe, 1985 (Halacaridae: Acari) from the Indian Ocean. J. Mar. Biol. Ass. India 37: 31–34.

Chilton C. 1883. On two marine mites. Trans. Proc. N. Z. Inst. Zool. 15: 190–192.

Coull B.C. 1972. Species diversity and faunal affinities of meiobenthic Copepoda in the deep sea. Mar. Biol. 14: 48–51.

Duxbury A.C., Duxbury A.B. and Sverdrup K.A. 2000. An introduction to the world's Oceans. 6th edition. McGraw Hill, Boston.

Gage J.D., Pearson M., Bilett D.S.M., Clark A.M., Jensen M., Paterson G.L.J. and Tyler P.A. 1984. Echinoderm zonation in the Rockall Trough (NE Atlantic). In: Keegen B.F. and O'Connor B.D.S. (eds), Proceedings of the Fifth International Echinoderm Conference, Galway. Balkema, Rotterdam, pp. 31–36.

Gage J.D. and Tyler P.A. 1991. Deep-Sea Biology: A Natural History of Organisms at the Deep-sea floor. Cambridge University Press, Cambridge.

Gosse P.H. 1855a. Notes on some new or little-known marine animals (II). Ann. Mag. Nat. Hist. 16(91): 27–36.
Gosse P.H. 1855b. Notes on some new or little-known marine animals (III). Ann. Mag. Nat. Hist. 16(91): 305–313.
Gray J.S. 2002. Species richness of marine soft sediments. Mar. Ecol. Prog. Ser. 244: 285–297.
Haedrich R.L., Rowe G.T. and Polloni P.T. 1975. Zonation and faunal composition of epibenthos populations on the continental slope south of New England. J. Mar. Res. 33: 191–212.
Hedgpeth J.W. 1963. Marine biogeography. In: Hedgpeth J.W. (ed.), Treatise on Marine Ecology and Palaeoecology. Vol. I (13): 359–382.
Hobday A.J. 2000. Persistence and transport of fauna on drifting kelp (*Macrocystis pyrifera* (L.) C. Agardh) rafts in the Southern California Bight. J. Exp. Mar. Biol. Ecol. 253: 75–96.
Hodge G. 1860. Contributions to the marine zoology of Seaham Harbour. 1. On a new marine mite (*Pachygnathus Seahami*). Trans. Tyneside Nat. Fld Cl. 4(3): 319.
Hodge G. 1863. Contributions to the marine zoology of Seaham Harbour. On some undescribed marine Acari. Trans. Tyneside Nat. Fld Cl. 5(4): 298–303.
Huang Z.G. 1994. Marine Species and their Distributions in China's Seas. China Ocean Press, Beijing.
Johnston G. 1836. Illustrations in British zoology. Mag. Nat. Hist. 9(63): 353–357.
Kirchner W.-P. 1969. Zur Biologie und Ökologie von *Halacarus basteri basteri* Johnson 1836 (Acari, Trombidiformes). Oecologia (Berl.) 3: 56–69.
Krantz G.W. 1970. *Agauopsis vidae*, a new species of Halacaridae (Acari: Prostigmata) from the Northern Adriatic Sea, with notes on its behavior. Archo Oceanogr. Limnol. 16: 247–261.
Laubier L. 1960. *Parhalixodes travei* n. g., n. sp., un nouvel Halixodinae (Halacariens) ectoparasite de nemerte en Méditerranée occidentale. Acarologia 2: 541–551.
Leppäkoski E., Gollasch S., Gruzka P., Ojaveer H., Olenin S. and Panov V. 2002. The Baltic—a sea of invaders. Can. J. Fish. Aquat. Sci. 59: 1175–1188.
Littler M.M., Littler D.S., Blair S.M. and Norbis J.N. 1985. Deepest known plant life discovered on an uncharted seamount. Science 227: 57–59.
Lohmann H. 1889. Die Unterfamilie der Halacaridae Murr. und die Meeresmilben der Ostsee. Zool. Jb. Syst. 4: 269–408.
Lohmann H. 1893. Die Halacarinen der Plankton-Expedition. Ergebn. Atlant. Ozean Plankton-Exped. Humboldt. Stift. 2: 11–95.
Lüning K. 1985. Meeresbotanik: Verbreitung, Ökophysiologie und Nutzung der marinen Makroalgen. Thieme Verlag, Stuttgart.
MacQuitty M. 1984. The feeding behaviour of two species of *Agauopsis* (Halacaroidea) from California. In: Griffith D.A. and Bowman C.E. (eds), Acarology, VI (1). John Wiley, New York, pp. 571–580.
Maguire B.Jr. 1963. The passive dispersal of small aquatic organisms and their colonization of isolated bodies of water. Ecol. Monogr. 33: 161–185.
Makkaveeva E.B. 1966. K biologii *Halacarellus basteri* (Jochnston) v Cernom More. In: Rasprедelnie bentosa i biologiya donnich zivotnich v juznach moryach, Akademia Nauk ukrainskoi SSR. Kiev. pp. 38–45.
Milbrink G. 1999. Distribution and dispersal capacity of the Ponto-Caspian tubificid oligochaete *Potamothrix heuscheri* (Bretscher, 1900) in Scandinavia. Hydrobiologia 406: 133–142.
Monniot F. 1964. Sur deux espèces du genre *Scaptognathus* présentes a Roscoff. Acarologia 6: 491–498.
Monniot F. 1967. Deux halacariens endopsammiques: *Halacarus anomalus* Trouessart 1894 et *Halacarus marcandrei* n. sp. Cah. Biol. Mar. 8: 89–98.
Morselli I. 1980. Su tre Acari Prostigmati di acque salmastre dell'alto Adriatico. Atti Soc. Tosc. Sci. Nat., Mem., Ser. B 87: 181–195.
Myers A.A. 1996. Species and generic gamma-scale diversity in shallow-water marine Amphipoda with particular reference to the Mediterranean. J. Mar. Biol. Ass. UK 76: 195–202.
Newell I.M. 1949. New genera and species of Halacaridae (Acari). Am. Mus. Novitat. 1411: 1–22.

Newell I.M. 1951. Further studies on Alaskan Halacaridae (Acari). Am. Mus. Novitat. 1536: 1–56.
Newell I.M. 1971. Halacaridae (Acari) collected during cruise 17 of the R/V Anton Bruun, in the southeastern Pacific Ocean. Anton Bruun Rep. 8: 3–58.
Newell I.M. 1984. Antarctic Halacaroidea. Antarctic Res. Ser. 40: 1–284.
Otto J.C. 1999. *Corallihalacarus chilcottensis*, a new genus and species of marine mites from the Coral Sea (Acarina: Halacaridae). Zool. Sci. 16: 839–843.
Otto J.C. 2000a. *Spongihalacarus longiscutus* n. gen., n. sp., a marine mite (Acari: Prostigmata: Halacaridae) associated with an alga-sponge symbiosis from the Great Barrier Reef Lagoon in Australia. Internat. J. Acarol. 26: 279–283.
Otto J.C. 2000b. *Xenohalacarus longirostris* n. gen., n. sp., a marine mite (Acari: Halacaridae: Lohmannellinae) from Chilcott Islet, Australia. Internat. J. Acarol. 26: 285–291.
Otto J. 2000c. Seven new species of *Arhodeoporus* (Acarina: Halacaridae) from the Great Barrier Reef and Coral Sea. Hydrobiologia 436: 1–16.
Otto J. 2001. Halacaridae of the Great Barrier Reef and Coral Sea: The *Copidognathus ornatus* group. (Acarina: Prostigmata: Halacaridae). Mem. Qd Mus 46: 717–731.
Packard A.C. 1871. On insects inhabiting salt water. Am. J. Sci. 3: 100–110.
Proches S. 2002. New species of Copidognathinae (Acari: Halacaridae) from southern Africa. J. Nat. Hist. 36: 999–1007.
Pugh P.J.A. and King P.E. 1986. Seasonality in some British intertidal Acari. J. Nat. Hist. 20: 653–666.
Rex M.A. 1981. Community structure in the deep-sea benthos. Ann. Rev. Ecol. Syst. 12: 331–353.
Rosenzweig M.L. 1997. Species Diversity in Space and Time, 3rd edn. Cambridge University Press, Cambridge, MA.
Sanders H.L. and Hessler R.R. 1969. Ecology of the deep-sea benthos. Science 163: 1419–1424.
Sarma A.L.N. and Chatterjee T. 1991. Occurence of *Copidognathus hartwigi* Bartsch (Halacaridae: Acari) from the Indian Ocean. J. Bombay Nat. Hist. Soc. 88: 300–302.
Sarma A.L.N. and Chatterjee T. 1993. Occurence of *Arhodeoporus bonairensis* (Viets, 1936) (Halacaridae: Acari) from Indian Ocean with zoogeographical remarks on genus *Arhodeoporus* Newell. J. Bombay Nat. Hist. Soc. 90: 417–422.
Siemer F. 1996. Untersuchungen zur Verteilung, zur Biologie und zum Lebenszyklus mariner Halacaridae (Prostigmata: Acari) im ästuarinen Felslitoral. University of Bremen, Ph. D. thesis, 165 pp.
Straarup J.-O. 1968. On the life cycles of halacarids (Acari) from the Öresund. Ophelia 5: 255–271.
Søvik G. 2002. *Ameronothrus lineatus* (Oribatida) on the high-Arctic archipelago of Svalbard: does synchronous moulting lead to an extended life cycle? In: Morales-Malacara J.B. and Rivas G. (eds), XI International Congress of Acarology. Program and Abstract Book, Merida, Yucatan, Mexico, pp. 279–280.
Viets K. 1936. Zoologische Ergebnisse einer Reise nach Bonaire, Curaçao und Aruba im Jahre 1930. No. 18. Halacariden aus Westindien. Zool. Jb. (Syst.) 67: 389–424.
Viets K. 1938. Eine merkwürdige, neue, in Tiefsee-Echiniden schmarotzende Halacaridengattung und -Art (Acari). Z. Parasitkde. 10: 210–216.
Viets K. 1939. Meeresmilben aus der Adria (Halacaridae und Hydrachnellae, Acari). Arch. Naturgesch. (N.F.) 8: 518–550.
Viets K.O. 1987. Die Milben des Süßwassers. 2. Katalog. Sonderb. Abh. naturw. Ver, Hamburg 1012 pp.
Wilhelmsen U. 1999. Rapid colonization of new habitats in the Wadden Sea by the ovoviviparous *Littorina saxatilis*. Helgolnder Meeresunters 52: 325–335.
Zezina O.N. 1997. Biogeography of the bathyal zone. Adv. Mar. Biol. 32: 389–426.

Experimental and Applied Acarology **34:** 59–65, 2004.

Traditional water mite fixatives and their compatibility with later DNA studies

ISABEL REY[1], BEATRIZ A. DORDA[1] and
ANTONIO G. VALDECASAS[2,3,*]
*[1]Museo Nacional de Ciencias Naturales (CSIC), Dpto. deColecciones, C/José Gutièrrez Abascal 2, 28006 Madrid, Spain; [2]Museo Nacional de Ciencias Naturales (CSIC), Dpto. Biodiversidad y Biología Evolutiva, C/José Gutiérrez Abascal 2, 28006 Madrid, Spain; [3]Authorship order determined at random; *Author for correspondence (e-mail: valdeca@mncn.csic.es)*

Key words: 16S, Angelier's fluid, CoI, Hydracharina, Koenike's fluid

Abstract. This work compares frozen water, 70% alcohol, Koenike's and Angelier's fluid as preservation media for water mites in terms of their eventual facilitation of DNA extraction and amplification. The time the mites spent in the fixatives ranged between 1 week and 25 years. Two molecular markers were amplified: 16S ribosomal DNA and CoI mitochondrial DNA. DNA was extractable and could be sequenced from specimens fixed in all the above media, although this generally became more difficult as time progressed. In the light of the known characteristics of these media, the results suggest Angelier's fluid to be the most practical, especially on long expeditions.

Introduction

In museums and taxonomic collections, water mites (Acari: Parasitengona: Hydrachnidia) are usually, but not always, mounted on microscope slides or preserved in Koenike's fluid (a mixture of acetic acid, glycerine and water; Barr 1973; Smith et al. 2001). Microscope slides are the preferred method, but very commonly the 'liquid collection' is larger.

Alcohol can be used to preserve most Acari, but this medium is unsuitable for Hydrachnidia, leaving specimens in poor condition for dissection and mounting on slides. Formol is even worse. A little used fixative, Angelier's fluid (mixture of water, chromic acid and acetic acid) can be used for whole benthic samples (Valdecasas and Baltanás 1989) or for water mites separated from such samples. It is particularly useful during field work since it can be concentrated, making it easier to transport. Preserving mites frozen in water is also a possibility.

To date, molecular research on the Hydrachnidia (Otto and Wilson 1999; Söller et al. 2001) has been performed with fresh material. No trials on the feasibility of extracting and sequencing DNA from mites preserved in the above fixatives have been performed. The aim of the present work was to determine the suitability of Koenike's fluid, Angelier's fluids, alcohol and frozen water in this respect.

Materials and methods

Water mite specimens were collected from the streams and ponds of the Sierra de Guadarrama (Madrid, Spain) during the spring of 2002. The stream/pond bottoms were kick-sampled with a 0.180-μm net and the collected material washed in the field through a 0.250-μm sieve. Samples were stored in a field fridge for transport to the laboratory. The other specimens analysed, which belonged to the Museo Nacional de Ciencias Naturales, Madrid, had been collected from the River Lozoya, Sierra de Guadarrama, and had been preserved in fixatives for up to 25 years.

The water mites collected in the field were separated from the sample debris in the laboratory with the help of a stereoscopic microscope and divided into four treatment groups. One was frozen at −20 °C, the second was stored in modified Koenike's fluid (45% water, 45%, glycerine, 10% glacial acid acetic; Cook 1974), the third was stored in 70% alcohol (analytical grade, diluted), and the fourth preserved in diluted Angelier's fluid (1% anhydrous chromic acid, 98%, water, 1% glacial acid acetic; Angelier 1953). The latter was made in a concentrated form and around 1 or 2 ml was used to preserve 100 ml benthic samples (Valdecasas and Baltanás 1989).

The possibility of extracting and sequencing the DNA of the mites was tested after different periods of preservation: 1 week after collection, approx. 3–4 months, 10 months, and after 15 and 25 years (specimens from the museum collection). The mites belonging to these last two samples had initially been stored in Angelier's fluid as whole benthic samples before their preservation in Koenike's fluid. Prior to analysis, all mites were identified at the genus level (*Eylais*, *Torrenticola*, *Sperchonopsis*, *Sperchon*, *Axonopsis*, *Kongsbergia*, *Aturus*, *Hygrobates* and *Protzia*) (Cook 1974).

Molecular methods

Single specimens were used for each analysis. Total genomic DNA was extracted using a modification of the CTAB (hexadecyltrimethylammonium bromide) method (Doyle and Doyle 1987). Specimens were briefly washed in distilled water and then crushed with a pipette tip against the wall of a 1.5 ml microcentrifuge tube containing 700 μL CTAB buffer (2% CTAB, 1.4 M NaCl, 0.2 M EDTA, 0.1 M Tris–HCl and 0.4% β-mercaptoethanol, pH 8.0) and 100 μg/ml proteinase K. The tubes were incubated overnight at 56 °C. DNA was extracted using phenol–chloroform–isoamyl alcohol followed by isopropanol precipitation. These samples were resuspended in TE buffer (10 mM Tris–HCl, 0.1 mM EDTA, pH 7.5) and purified using the silica method with some modifications (Boom et al. 1990). A fragment of mitochondrial (mt) 16S rDNA and CoI was amplified using 0.4 μM of primer 16Sar (5′-CGCCTGTTTATCAAAAACAT-3′), 16Sbr (5′-CCGGTCTGAACTCA GATCACTG-3′) (Palumbi et al. 1991), CoIH (5′-TCAGGGTGAC

CAAAAAATCA-3′), and CoIL (5′-GGTCAACAAATCATAAAGA TATTGG-3′) (Folmer et al. 1994). Five μl of the CTAB-extracted DNA solution was used as a template. The other PCR agents were 75 mM Tris–HCl (pH 9.0), 2 mM $MgCl_2$, 50 mM KCl, 20 mM $(NH_4)_2SO_4$, 10 mM dNTPs, 0.02% BSA and 0.625 units Taq DNA polymerase (Biotools) in a final volume of 25 μl.

The PCR involved an initial denaturation step at 94 °C for 5 min, followed by 40 cycles of 94 °C for 30 s, 51 °C for 45 s, and 72 °C for 45 s, and a final extension step of 72 °C for 10 min. The end products were stored at 4 °C before separation on 2% agarose gels in 1× TAE buffer (40 mM Tris–acetate, pH 8.0 and 1 mM EDTA). Bands were stained with ethidium bromide and photographed under UV light. The presence or absence of DNA fragments was scored against a molecular mass marker. Several end products were then purified and concentrated using Kit Bioclean columns (Biotools) for sequencing. Each strand was sequenced using 'Big Dye Terminator' (Applied Biosystems, Inc.; ABI) sequencing reactions for each primer, and analysed using an ABI Capillary 3700 Genetic Analyzer.

All DNA extracts were given a unique code and stored at −80 °C as part of the Tissue and DNA Collection of the Museo Nacional de Ciencias Naturales, Madrid (available from the authors upon request).

Results and discussion

DNA was extracted from specimens stored in all the tested media. Figure 1 shows an agarose gel with fragments of the amplified CoI gene. Sequences of the amplified fragments were deposited in the GenBank database with accession numbers AY393896 *Eylais* sp. (MNCN/ADN: 10539), AY393897 *Lebertia* sp. (MNCN/ADN: 10562), AY393898 *Lebertia* sp. (MNCN/ADN: 10587), AY393899 *Sperchonopsis* sp. (MNCN/ADN: 10541), AY393900 *Torrenticola* sp. (MNCN/ADN: 10547) and AY392745 *Atractides* sp. (MNCN/ADN: 10560).

Table 1 and Figure 2 show the numbers of specimens from which DNA was successfully extracted for each method and duration of preservation. As expected, DNA was most easily extracted (100% success) from the organisms frozen in water (Dessauer and Hafner 1984; Prendini et al. 2001). No results are available for mites stored in this medium for more than 4 months.

DNA was less regularly extractable from alcohol-preserved mites. Extractions were successful from 15 out of 18 specimens at 4 months, from only two out of five at 10 months, and from just one out of five at 15 years. Many empirical data on DNA extraction from alcohol-preserved organisms are available, and the results agree with those expected, that is, alcohol preservation is less successful than cryoconservation – yet it may yield satisfactory quantity and quality of DNA for PCR (Oliveira et al. 2002). As a matter of fact, alcohol is the preferred preservation fluid for zoological material after cryoconservation (Greer et al. 1991; Prendini et al. 2001).

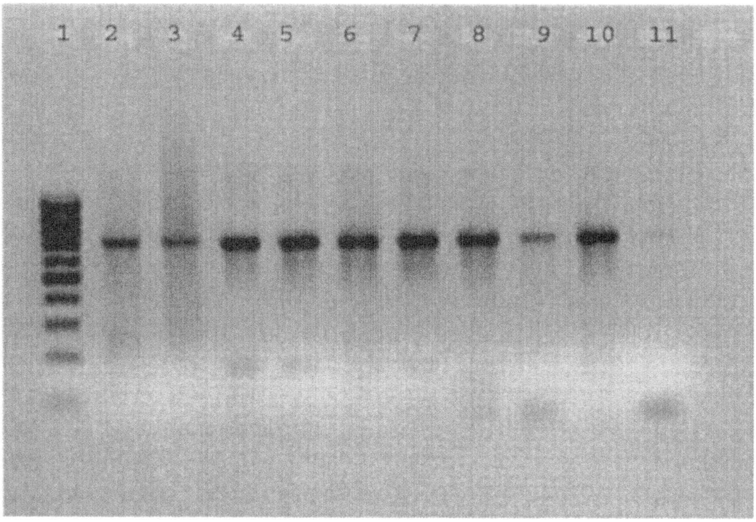

Figure 1. Two percent agarose gel stained with ethidium bromide and photographed under UV light. Note the PCR bands of the CoI gene fragment. 1, 100 bp ladder; 2–3, *Eylais* sp. frozen for 1 week; 4–6, *Sperchonopsis* sp. frozen for 4 months; 7–9, *Atractides* sp. after 4 months in 70% alcohol; 10–11, *Lebertia* sp. after 4 months in Koenike's fluid.

Eight out of 17 mites stored in Koenike's fluid for 4 months yielded DNA, but none was obtained from those that had been in this fluid for 10 months. Surprisingly, however, DNA was extracted successfully from two out of five mites stored for 25 years. With respect to Angelier's fluid, DNA was extracted from 11 out of 15 mites at 4 months, from two out of five mites at 10 months and, surprisingly again, from four out of five after 25 years of storage. To our knowledge, this is the first time that DNA was extracted from water mite specimens stored in Koenike's or Angelier's fixatives. In similar studies, Dillon et al. (1996) and Quicke et al. (1999) compared the suitability of 100% ethanol, ethylene glycol, formalin and storage at −80 °C for preserving Hymenoptera specimens to be used in molecular studies. However, neither group tested the suitability of Koenike's nor Angelier's fluid.

For the alcohol and Angelier's fluid methods, the sooner the organisms were analysed after collection and fixation, the more easily their DNA was extracted and sequenced. As time passed, the number of unsuccessful extraction attempts increased. Paradoxical results were obtained with Koenike's fluid. No DNA could be extracted after 10 months, although two specimens provided DNA after 25 years. Further analysis is necessary to determine whether this is an effect of the taxa involved. However, it is known that the amount of Taq polymerase inhibitors (mainly polysaccharides and pigments; Doyle and Doyle 1987) introduced into the PCR environment by the sample under test can vary greatly depending on its physiological state at the time of capture. For example, crustaceans produce anti-freezing proteins at certain times of year to

Table 1. Extraction success (%) obtained with the different storage media: duration of storage and different taxa.

Storage medium	Time	Taxa (number of specimens)	% Success
−20 °C (Water)	1 week	*Eylais* sp. (5)	100
	4 months	*Torrenticola* sp. (12)	100
		Sperchonopsis sp. (3)	100
		Atractides sp. (2)	100
		Hygrobates sp. (1)	100
70% Alcohol	4 months	*Torrenticola* sp. (11)	90
		Atractides sp. (3)	100
		Lebertia sp. (2)	100
		Aturus sp. (1)	0
		Sperchon sp. (1)	0
	10 months	*Torrenticola* sp. (5)	40
	15 years	*Axonopsis* sp. (3)	33
		Torrenticola sp. (1)	0
		Kongsbergia sp. (1)	0
Koenike's	4 months	*Torrenticola* sp. (9)	33
		Atractides sp. (4)	25
		Lebertia sp. (2)	100
		Protzia sp. (1)	100
		not identified (1)	100
	10 months	*Torrenticola* sp. (5)	0
	25 years	*Torrenticola* sp. (5)	40
Angelier's	4 months	*Torrenticola* sp. (11)	63
		Atractides sp. (2)	100
		Hygrobates sp. (1)	100
		Aturus sp. (1)	100
	10 months	*Torrenticola* sp. (3)	66
		Lebertia sp. (1)	0
		Aturus sp. (1)	0
	25 years	*Lebertia* sp. (3)	100
		Torrenticola sp. (1)	0
		Hygrobates sp. (1)	100

help them endure very low temperatures, and these can inhibit the PCR (Branquart et al. 1996).

It is important to note that the primers used in this work amplify DNA segments longer than 700 bp. However, fragments smaller than 600 bp were obtained from 25-year-old samples, probably because of shearing. When working with ‘ancient DNA’ (samples older than 10 years can be considered ancient if they have been subjected to shearing–inducing temperature changes), it is better to use primers for shorter sequences. Further research is necessary to determine whether the DNA of water mites preserved over long periods of time is better amplified with the techniques used for ancient DNA.

These empirical results could be of great practical use to water mite researchers. After mites are collected and preserved in Angelier’s or Koenike’s

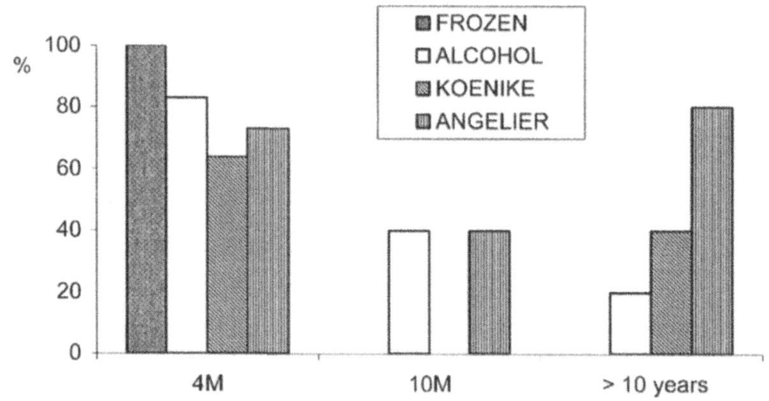

Figure 2. Percentage of specimens whose DNA was successfully extracted after different durations of storage (M, month).

Table 2. Cost, weight and other variables associated with the use of the different fixatives on long expeditions.

Fixative	Cost	Weight	Safety concerns	Remarks
Freezing	High	Medium	None	Difficult to perform
Alcohol	High	High	Potentially dangerous	Problems of acquisition and transport
Koenike's fluid	Low	High	None	Large amounts of fixative required per whole sample
Angelier's fluid	Low	Low	None	Leaks easily from non-airtight containers

fluid, there appears to be a 'safe period' of at least 3 months during which their DNA can be successfully analysed. Table 2 compares the cost and weight etc. of the tested fixatives. These data, plus the experimental results obtained, suggest that Angelier's fluid is the most practical fixative for water mites. DNA can be extracted from samples even after long periods of storage and, unlike alcohol, morphological studies are not compromised. Angelier's fluid is also cheap, can be concentrated for transport and is easy to use, making it an optimum medium for use on long sampling trips (Valdecasas and Baltanás 1989).

Acknowledgements

Ana Camacho gave her time generously to improve this paper. Adrian Burton kindly revised the English version. This work was funded by the following projects: EVK2-CT-2001-00121 (Pascalis) and REN2000-2040GLO, and by the resources of the Tissues and DNA Collection of the MNCN.

References

Angelier E. 1953. Recherches écologiques et biogéographiques sur la faune des sabeles sumergés. Arch. Zool. Exp. Gen. 90: 37–161.

Barr D. 1973. Methods for the Collection, Preservation and Study of Water Mites (Acari: Parasitengona). Life Sciences Miscellaneous Publications, Royal Ontario Museum, pp. 1–28.

Boom R., Sol C.J., Salimans M.M., Jansen C.L., Wertheim-van Dillen P.M. and der Noordaa J. 1990. Rapid and simple method ford purification of nucleic acids. J. Clin. Microbiol. 28(3): 495–503.

Branquart E., Hemptinne J.L. and Gaspar C. 1996. Adaptation au froid des invertébrés et cryoconservation des œufs d'insectes aphidiphages. Cah. Etud. Rech. Franc. Agric. 5(5): 353–364.

Cook D.R. 1974. Water Mite Genera and Subgenera. Memoirs of the American Entomological Institute, Vol. 21, pp. 1–860.

Dessauers H.C. and Hafner M.S. 1984. Collections of Frozen Tissues: Value, Management, Field and Laboratory Procedures and Directory of Existing Collections. Association of Systematics Collections, University of Kansas Press, Lawrence, KS.

Dillon N., Austin A.D. and Bartowsky E. 1996. Comparison of preservation techniques for DNA extraction from hymenopterous insects. Insect Mol. Biol. 5(1): 21–4.

Doyle J.J. and Doyle J.L. 1987. A rapid DNA isolation procedure for small quantities of fresh leaf tissue. Phytochem. Bull. 19: 11–15.

Folmer O., Black M., Hoeh W., Lutz R. and Vrijenhoek R. 1994. DNA primers for amplification of mitochondrial cytochrome *c* oxidase subunit I from diverse metazoan invertebrates. Mol. Mar. Biol. Biotechnol. 3(5): 294–299.

Greer C.E., Peterson S.L., Kiviat N.B. and Manos M.M. 1991. PCR Amplification from paraffin-embedded tissues: recommendations on fixatives for long-term storage and prospective studies. PCR Meth. Appl. 1: 46–50.

Oliveira C.M., Fungaro M.H.P., Camargo L.E.A. and Lopes J.R.S. 2002. Anàlise Comparativa da Estabilidade do DNA de *Dalbulus maidis*(DeLong & Wolcott) (Hemiptera: Cicadellidae) sob Diferentes Métodos de Preservaçao Para Uso em RAPD-PCR. Neotrop. Entomol. 31: 225–231.

Otto J.C. and Wilson K.J. 1999. Assessment of the usefulness of ribosomal 18S and mitochondrial COI sequences in Prostigmata phylogeny. In: Halliday R.B., Walter D.E., Proctor H.C., Norton R.A. and Colloff M.J. (eds), Acarology. Proceedings of the 10th International Congress. CSIRO Publishing, pp. 100–109.

Palumbi S.R., Martin A.P., Romano S., McMillan W.O., Stice L. and Grabowski G. 1991. The Simple Fool's Guide to PCR. Special Publ. Dept. Zoology, University of Hawaii, Honolulu.

Prendini L., Hanner R. and DeSalle R. 2001. Obtaining, Storing and Archiving Specimens and Tissue Samples for use in Molecular Studies. http://research.amnh.org/amcc/papers.html.

Quicke D.L.J., Belshaw R. and Lopez-Vaamonde C. 1999. Preservation of hymenopteran specimens for subsequent molecular and morphological study. Zool. Script. 28: 261–267.

Smith I.M., Cook D.R. and Smith B.P. 2001. Water mites (Hydrachnidia) and other arachnids. In: Thorp J.H. and Covich A.P. (eds), Ecology and Classification of North American Invertebrate, 2nd ed. Academic Press, San Diego, pp. 551–659.

Söller R., Wohltmann A., Witte H. and Blohm D. 2001. Phylogenetic relationships within terrestrial mites (Acari: Prostigmata, Parasitengona) inferred from comparative DNA sequence analysis of the Mitochondrial Cytochrome Oxidase subunit I gene. Mol. Phylogenet. Evol. 18: 47–53.

Valdecasas A.G. and Baltanás A. 1989. A note on the use of Angelier's fluid for freshwater invertebrates. Arch. Hydrobiol. 115: 313–316.

Experimental and Applied Acarology **34:** 67–77, 2004.

Random amplified polymorphic DNA analysis of kinship within host-associated populations of the symbiotic water mite *Unionicola foili* (Acari: Unionicolidae)

DALE D. EDWARDS*, DANIEL E. DEATHERAGE and BRIAN R. ERNSTING

*Department of Biology, University of Evansville, 1800 Lincoln Ave., Evansville, IN 47722, USA; *Author for correspondence (e-mail: de3@evansville.edu; phone: +1-812-479-2645; fax: +1-812-488-1039)*

Key words: Kinship, Random amplified polymorphic DNA, *Unionicola foili*

Abstract. Kinship relations within populations of unionicolid water mites are not well known, owing to their complex life cycles and the fact that interactions between active and resting stages for some species are transitory. A number of species of unionicolid water mites are, however, obligate symbionts of freshwater mussels and spend most of their life cycle in association with these hosts. Among these species of mites, parents and offspring are more likely to co-occur and thus provide opportunities to address questions related to the structure of the mating system. The present study employs random amplified polymorphic DNA (RAPD) analysis to address kinship within populations of *Unionicola foili* living in symbiotic association with the host mussel *Utterbackia imbecillis*. DNA was amplified from adult mites and a representative number of eggs or larvae (n = 20–30) that were removed from mussels collected on three separate occasions (July, November, and March) over a 12-month period. Parsimony analyses of the molecular data for adults and progeny collected from mussels during July, November, and March revealed distinct groupings, that for the most part, corresponded to mites collected from each of the sampling periods. Many of the genetic markers obtained for male and female *U. foili* were not evident among the larvae or eggs, suggesting that adults obtained from a host mussel at the time of collection were not the parents of a majority of the progeny. However, female mites and eggs collected from mussels during March and November shared more markers than did females and progeny examined during July. Furthermore, many offspring in the July sampling period were found to have one or more parents absent from the sampled population. Overall, RAPD profiling appears to have limited usage in determining kinship within populations of *U. foili*, due to its recruitment patterns, and the relatively large number of adults and progeny per mussel. It may, however, prove to be a useful method for assessing genetic relatedness among unionicolid mussel-mites that have substantially lower population densities.

Introduction

The life cycle of water mites (Acarina: Hydrachnidia) includes egg, larva, protonymph, deutonymph, tritonymph, and adult. Larvae of most species undergo a brief parasitic association with aerial insects (Smith and Oliver 1986) and subsequently must return to the aquatic environment to complete their life cycle. The adults and deutonymphs of most water mite species are free-living predators (Gledhill 1985). There are, however, species from the family Pionidae and Unionicolidae that are symbiotic with freshwater gastropods, mussels, and sponges (Mitchell 1955; Hevers 1980).

Water mites of the genus *Unionicola* (Haldeman) are common symbionts of freshwater mussels (Vidrine 1996). Some species are free-living predators as nymphs and adults, depending upon the host only for sites for oviposition and post-larval resting stages, while others are obligate symbionts of their host (Mitchell 1955). Adult mites presumably mate within the confines of the mantle cavity (Hevers 1978), with females subsequently depositing fertilized eggs in host gill tissues. Larvae emerge from the gills during spring and summer, leave the host mussel, and establish a parasitic association with pupae of the family Chironomidae (Jones 1978). When these insects molt, the larvae are dragged through the pupal exuvium and begin to engorge on host hemolymph (Böttger 1972; Hevers 1980; Gledhill 1985). Larvae that are attached to a host insect after it becomes aerial must return to the aquatic environment and re-enter a host mussel to complete their life cycle.

The North American unionicolid water mite *Unionicola foili* (Edwards and Vidrine) is an obligate symbiont that commonly occurs in permanent association with the freshwater mussel *Utterbackia imbecillis* (Say) (Vidrine 1996). Although the mite *Unionicola formosa* (Dana and Whelpley) traditionally has been described from freshwater mussels of the genera *Utterbackia* (Baker) and *Pyganodon* (Crosse and Fischer) (Vidrine 1996), Edwards and Vidrine (1994) have separated *U. formosa* into two species, *U. formosa* from mussels of the genus *Pyganodon* and *U. foili* from the genus *Utterbackia*. *Unionicola foili* is among the best studied water mites in North America, with published studies on its population dynamics (Dimock 1985; Edwards and Dimock 1988), genetic diversity (Edwards and Dimock 1997), patterns of host specificity (LaRochelle and Dimock 1981) and zoogeography (Dobson 1966; Vidrine 1996). Population studies indicate that *U. foili* exhibits a female-biased sex-ratio, with nearly 100% of host mussels harboring a single male and 40 or more females (Dimock 1985). Experimental and field evidence suggest that intrasexual aggression and territoriality by males are, in part, responsible for this distribution (Dimock 1983). The territorial behavior displayed by males constitutes a female-defense polygynous mating system. Because male *U. foili* have exclusive access to numerous females within a host's mantle cavity, successfully defending that territory could potentially increase a male's fitness.

With the exception of the aforementioned studies, nothing else is known about the structure and dynamics of the mating system of *U. foili*. This unionicolid mussel-mite should, however, be well suited for addressing questions related to kinship, given its obligate relationship with mussels and the fact that adults, nymphs, and post-larval resting stages live concurrently in the same host. A number of species of unionicolid water mites are free-living predators as nymphs and adults, depending on a host only for oviposition and post-larval development (Mitchell 1955; Hevers 1980).

Over the past 10 years, molecular genetic approaches have become increasingly important to studies in behavioral and population ecology. In particular, the ability to determine the genetic structure of populations using DNA fingerprinting technologies has enhanced our understanding of animal mating systems by permitting analyses of kinship relationships (Ross 2001). Unionicolid water mites present two important challenges to this type of analysis: (1) their genome is almost completely uncharacterized, and (2) individuals are small, limiting the amount of DNA available from a single individual. These constraints require that any genetic study involve amplification of template DNA without relying on extensive DNA sequence information. The use of random amplified polymorphic DNA (RAPD) markers addresses both of these concerns (Hadrys et al. 1992). The present study uses RAPD markers to address parentage among *U. foili* occurring in symbiotic association with its host mussel *U. imbecillis.*

Materials and methods

Study animals

Unionicola foili used for this study were obtained from a population of *Utterbackia imbecillis* collected from a 4-ha pond located in Perry Co., Indiana, U.S.A. (37° 56′N, 86° 43′W). A total of six mussels was collected, two each on three occasions over a 12-month period: summer (July 2002), autumn (November 2002), and winter (March 2003). Mussels were collected over different seasons because of the dynamic nature of displacement and recruitment of adult mites into the population (Dimock 1985). In the laboratory, all adults and a representative number of eggs or larvae (n = 20–30) were removed from a host mussel, washed several times in deionized water, and held in this medium for approximately 12 h. Mites were placed individually in 1.5 ml microcentrifuge tubes and stored at −70 °C, awaiting DNA extraction.

DNA extraction

Disposable pestles for grinding individuals were made by briefly heating 1000 μl pipette tips in a flame and pressing them into the bottom of a 1.5 ml centrifuge tube mold. The resulting pestles conformed to the shape of the storage tube, and allowed mites to be ground without being removed from the tube. After grinding, genomic DNA from individual mites was extracted using commercially available affinity methods: Geneclean™ or Qiagen DNeasy™ Tissue extraction kits.

PCR and electrophoresis

DNA amplification was carried out using a modification of the method of Williams et al. (1993). Twenty-five microliters reactions contained 2.5 units of

Taq DNA polymerase (Promega; Catalog number M1665), 10 mM Tris–HCl, pH 9.0, 50 mM KCl, 0.1% Triton X-100, 2 mM $MgCl_2$, 0.001% gelatin, 150 μM each dATP, dCTP, dGTP, and dTTP, 5 pmols of a single 10-base oligonucleotide primer, and 2 μl of template DNA (approximately 50 ng). The polymerase chain reaction used 45 cycles of denaturation at 94 °C for 1 min, annealing at 34 °C for 1 min, and extension at 72 °C for 2 min. PCR products were separated on 0.9% agarose gels and detected using ethidium bromide staining.

RAPD analysis

Parentage was determined by visual inspection of the markers present for each amplification reaction and assigning each marker as either present or absent in each individual. General relatedness among *U. foili* collected from mussels for each of the sampling periods was determined by parsimony analysis on genetic markers obtained from adult mites and offspring for three primers. The original data matrix is available from the corresponding author upon request. Parsimony analysis was done using heuristic search with nearest-neighbor interchange (NNI) branch swapping algorithm and MAXTREES set to 7200. Analysis was conducted using PAUP*4.0b10 (Swofford 2002). The shortest trees that were recovered were pooled and used to generate a majority rule consensus tree with majority-rule option set at 50%. Statistical support for the most parsimonious trees was assessed with bootstrap analysis on the original data matrix by generating 1000 pseudoreplicates.

Results

Individuals analyzed

All mussels from which mites were collected harbored one male and had a mean of 12.3 ± 1.7 (SE, range = 6–22) females/host. For mussels collected in July, offspring consisted mostly of larvae. In the autumn and winter sampling periods, there were no larvae present and all of the progeny from a mussel were eggs.

Preliminary data

To assess the suitability of RAPD profiling for determining parentage among *U. foili*, twelve 10-base oligonucleotide primers (Primer Kit A, Operon Technologies Inc.; Qiagen) were used individually to amplify DNA fragments for adult mites obtained from an individual mussel. Of those 12 primers, three (OPA03 [5′-AGTCAGCCAC], OPA09 [5′-GGGTAACGCC], and OPA11

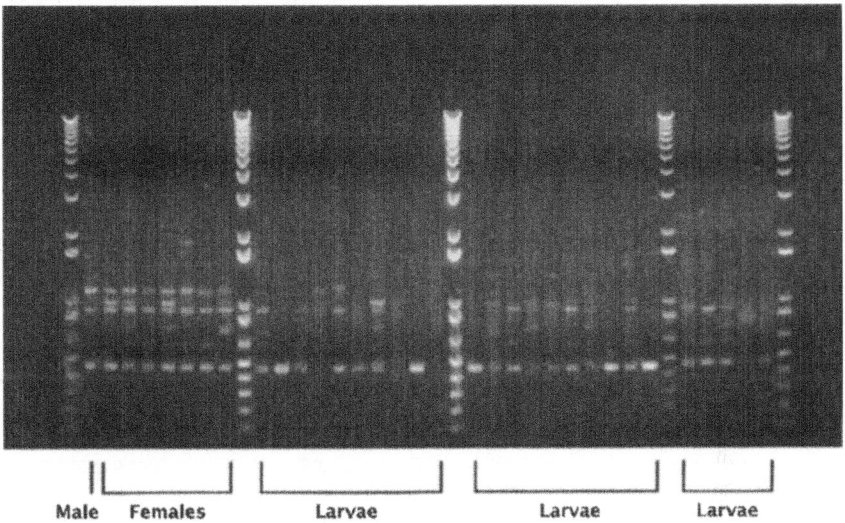

Figure 1. Agarose gel electrophoresis of RAPD products from *Unionicola foili* using primer OPA3. DNA was extracted from individuals isolated from a host mussel during the July sampling period. Five lanes contain molecular size standards. The remaining lanes contain RAPD products from all of the individuals isolated from this population. Male, females, and larvae are indicated at the bottom of the gel. Products are visualized by ethidium bromide staining.

[5′-CAATCGCCGT]) yielded a mixture of some PCR products that were present in most or all individuals and some that were present in a subset of individuals.

When RAPD amplifications were repeated on different days, using the same template DNA samples and the same primer, the number and molecular size of the markers were the same for most individuals. The reproducibility of RAPD markers was determined by comparing at least two profiles from the same population of *U. foili* for primers OPA03, OPA09, and OPA11. Where differences did occur, the bands that were not reproduced were very faint. For our determination of kinship relations, we analyzed the inheritance of RAPD markers that exhibited the same profiles when RAPD–PCR was repeated. All tissues including eggs yielded sufficient DNA for amplification using three primers. RAPD products for one primer from a population of mites removed from a host mussel collected in July are shown in Figure 1.

Parsimony analyses of RAPD profiles

A 50% majority rule consensus tree using RAPD profiles of individual *U. foili* from a representative mussel from July, November, and March is presented in Figure 2. The data set revealed 67 parsimony informative-characters. Heuristic searches identified 3280 equally parsimonious trees with a tree length of 322

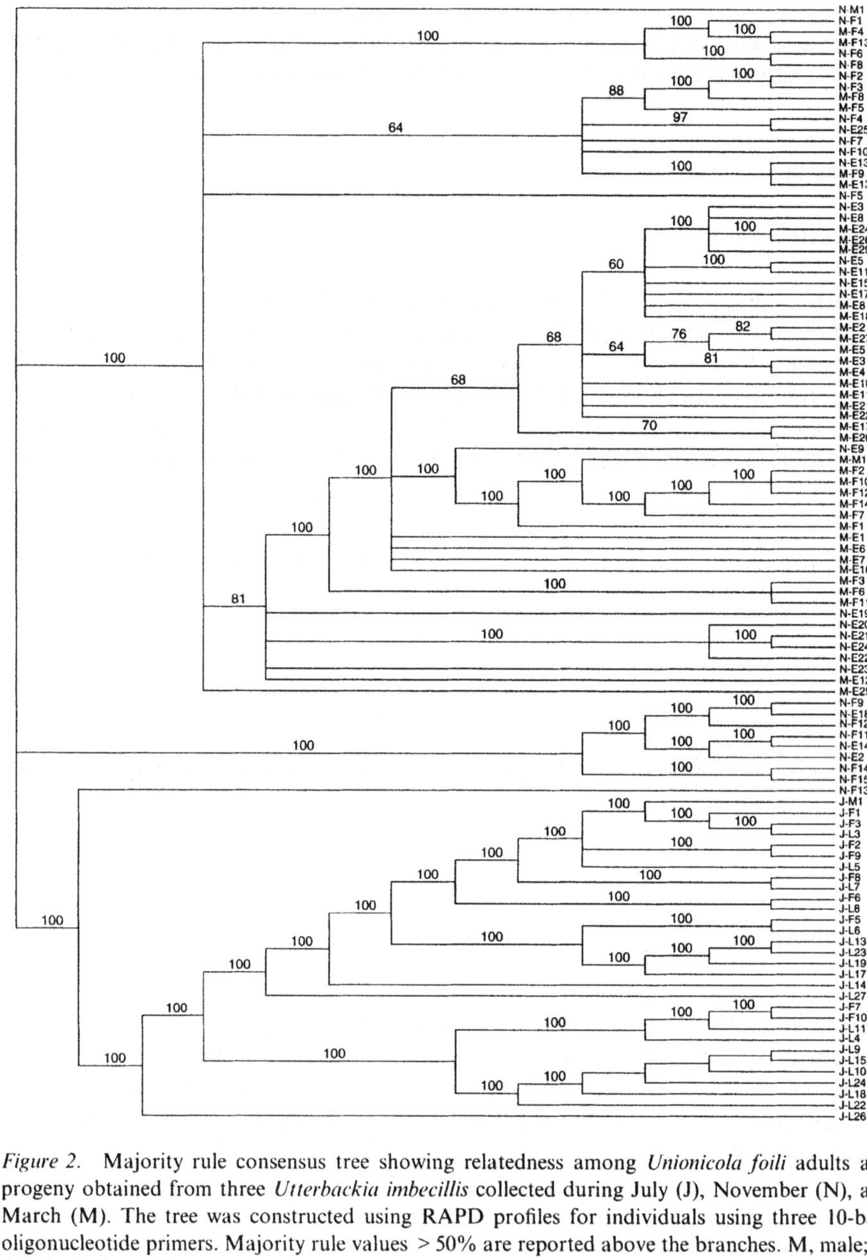

Figure 2. Majority rule consensus tree showing relatedness among *Unionicola foili* adults and progeny obtained from three *Utterbackia imbecillis* collected during July (J), November (N), and March (M). The tree was constructed using RAPD profiles for individuals using three 10-base oligonucleotide primers. Majority rule values > 50% are reported above the branches. M, male; F, female; L, larva; E, egg.

steps. Bootstrap analysis did not provide statistical support for most of the relationships presented in the 50% majority rule consensus tree. The majority rule consensus tree revealed distinct groupings, which with some exceptions, corresponded to mites obtained for each of the sampling periods.

RAPD profiles of U. foili *within host mussels*

Many of the genetic markers obtained for adult *U. foili* were not evident among larvae or eggs, suggesting that the male and females examined in a host mussel at the time of collection were not the parents of a majority of the progeny. However, adult mites and eggs collected from one mussel in March and from one mussel in November shared more markers than did adults and larvae obtained from other mussels. This became particularly evident when we examined which offspring (either larvae or eggs) from a single mussel contained one or more bands that were not present in any of the adults in the same host mussel (Table 1). Any band that was present in an offspring but absent in all adults excluded this individual offspring from being the progeny of the extant parents. In one representative June population, 90% of the offspring had one or both parents absent from the host's mantle cavity, with the excluded

Table 1. Number of exclusions for each offspring in three sampled populations of *U. foili*.

June		November		March	
Individual	Number of Exclusions	Individual	Number of Exclusions	Individual	Number of Exclusions
J-L3	1	N-E2	1	M-E1	0
J-L4	3	N-E3	1	M-E2	0
J-L5	3	N-E5	0	M-E3	0
J-L6	1	N-E8	1	M-E4	0
J-L7	0	N-E9	1	M-E5	0
J-L8	0	N-E11	0	M-E6	0
J-L9	6	N-E13	0	M-E7	0
J-L10	7	N-E14	0	M-E8	0
J-L11	7	N-E15	0	M-E9	0
J-L13	4	N-E17	0	M-E10	0
J-L14	1	N-E18	0	M-E11	0
J-L15	6	N-E19	0	M-E12	0
J-L17	3	N-E20	0	M-E13	0
J-L18	1	N-E21	0	M-E16	0
J-L19	1	N-E22	0	M-E17	0
J-L22	2	N-E23	0	M-E18	0
J-L23	5	N-E24	0	M-E20	1
J-L24	1	N-E25	0	M-E21	0
J-L26	3			M-E22	1
J-L27	1			M-E24	0
				M-E25	1
				M-E26	0
				M-E27	0
				M-E29	0
90% of offspring excluded		22% of offspring excluded		13% of offspring excluded	

Offspring were considered excluded if a marker present in an offspring was absent in all of the adults. J, July sampling period; N, November sampling period; M, March sampling period. L, larvae; E, eggs.

offspring having an average of 2.7 bands not present in any of the adults. During both November and March, one of the two sampled populations showed a much higher degree of relatedness, with 22% (November), and 13% (March) of offspring excluded in this way. Furthermore, during March and November, those offspring that were excluded all displayed a single band that was not present in the adults, in contrast to the multiple exclusions seen among summer populations.

Discussion

The results of the RAPD analyses indicate complex kinship relationships among the symbiotic water mite *U. foili*. Overall, a large percentage of RAPD markers found among female *U. foili* were not shared by larvae or eggs recovered from the same host mussel. In fact, most larvae and eggs were more closely related to each other than they were to adults collected from the same host.

With few exceptions, the results of the RAPD analysis showed that males had markers that were absent from most or all of the offspring. These results suggest that males that fathered the offspring may have been displaced by intruding males, perhaps through the intraspecific aggressive encounters that are characteristic of this species (Dimock 1983). The RAPD data also suggest that displacement of males may occur rather often. To determine the rate of turnover of resident males would require a more comprehensive sampling program than presented in this study and a better understanding of frequency of aggressive encounters between male mites in the field.

Despite overall low levels of similarity, there were higher degrees of genetic relatedness among female *U. foili* and immature stages that were collected during autumn (November) and winter (March) when compared to the summer (July) collecting period. This pattern of unrelated offspring and adults was observed in six populations of mites collected from mussels during the summer months, as well as in one of the two populations sampled in November, and one of the two populations sampled in March (data not shown). It was only during November and March that we observed any populations where the adults and offspring were closely related. In one of the two populations from each of these months, the majority of offspring present appear to be the offspring of the extant parents, and the parsimony analysis separates these individuals into closely related clusters. In general, the results of this study are consistent with seasonal changes in the population structure that have been reported for this mussel-mite. For example, Dimock (1985) reported annual minimum density of female *U. foili* per mussel in June and July but also found a higher proportion of small females in mussels during May and June. Dimock (1985) suggested that late spring–early summer represented a major period of recruitment into the adult population, with overwintering nymphs replacing the previous generation of females. The lack of genetic relatedness among females and larvae from the summer sampling period may thus not be surprising if larvae emerging from the gills of host mussels during summer are the progeny of females that oviposited during winter and were subsequently replaced by nymphs that were recruiting into the population.

An increase in the genetic similarity between female *U. foili* and eggs collected during November is consistent with the data pertaining to oviposition and the size-frequency distribution of females at this time. Dimock (1985) reported an increase in the density of female mites among *U. imbecillis* in November and also observed an increase both in the number of eggs per female and occurrence of mite eggs in host gills during this period. It is possible that females recovered from mussels during November represent individuals that were recruited in to the host population over the summer and that these mites were responsible for the increase in the number of eggs in the gills of mussels during winter. Similarly, the relatively high degree of genetic relatedness among female *U. foili* and eggs examined from mussels collected in March correspond to the concomitant maxima in the density of eggs per female and density of eggs in the gill tissue of *U. imbecillis* (Dimock 1985). Gravid females present in mussels during late winter–early spring are likely responsible for the large increase in the density of eggs in the host's gill during this period.

The RAPD methods reported in this study represent the first time that molecular markers have been used to characterize aspects of the mating system of water mites. DNA extraction from isolated individuals yielded sufficient DNA for multiple RAPD assays even from individuals in early stages of development, and RAPD profiles of eggs were as reproducible as those generated from adults. Initially, we were concerned that DNA extractions from early stages of a small species living inside the host would introduce unacceptable levels of contamination with host tissue, leading to RAPD products that indicative of the genetic state of the host rather than the symbiotic mites. Overnight incubation after extraction from the host along with extensive washing and removal of any residual liquid that might contain host tissue were carried out with each individual to decrease the ratio of host DNA to mite DNA in the extracted samples. The reproducible variation between individuals suggests that contamination with host DNA is not a serious concern in this system.

There are, however, a number of reasons why the RAPD data presented in this study should be interpreted with some degree of caution. First, RAPD profiles were generated for mites from only two host mussels for each of the sampling periods. A larger number of host-associated populations of *U. foili* from each of the sampling periods will need to be examined before more definitive conclusions regarding kinship relations for this species can be made. Interestingly, we have RAPD profiles for individual *U. foili* from three additional mussels collected during July and the genetic relatedness among mites from these mussels is consistent with the RAPD data obtained for mites from the two host mussels collected in July presented in this study. Second, there are contrasting differences in the degree of genetic relatedness among individuals from the two host mussels collected during both the November and March sampling periods, with one host-associated population of mites exhibiting a relatively high degree of similarity and the other showing relatively low degree of relatedness. Third, the bootstrap analysis did not provide support for the relationships generated by the 50% majority consensus trees. Finally, the anonymity of RAPD products, and our electrophoretic separation on the basis of product size raise the possibility of occasionally combining two non-identical markers that happen to be about the same molecular size as identical (Backeljau et al. 1995). This type of noise should, however, be rare, given the relatively small number of bands produced by any single RAPD primer, and the ability demonstrated by the wide separation of marker bands in the size range of the RAPD products in this assay to resolve size differences as small as a few percent. Furthermore, since the parsimony analysis is based on the shared inheritance of a large number of characters, it should be robust to the possibility of this type of occasional misidentification of non-identical characters as identical.

RAPD profiling has been useful in analyzing breeding system properties for an array of animal species, including ants, bees, and spiders (see Ross 2001 for a review), but it appears to have limited usage in addressing kinship within populations of the symbiotic water mite *U. foili*. The population dynamics of *U. foili*, coupled with extraordinary large numbers of eggs (maximum density of eggs/cm^2 = +300; Dimock 1985) and larvae (maximum density of larvae/cm^2 = +90; Dimock 1985) in the gills of host mussels, makes it particularly difficult to elucidate the precise nature of the genetic relationships between adults and progeny for this species. Future studies may, in part, circumvent these issues by removing eggs from gravid females and comparing the RAPD profiles of these eggs to those of their mothers and the resident male. Paternity relations among *U. foili* could be examined by removing a resident male and gravid females from a host mussel, introducing a new male, and determining whether the newly established male fathered the remaining females' offspring. Paternity in *U. foili* could also be addressed by inducing females to abort their eggs and determining whether the resident male was the sole father of future offspring. RAPD may prove to be a useful method for assessing the outcome of sperm competition for any species of water mite that will mate and develop eggs under laboratory conditions.

Acknowledgements

We thank Eric Janson, Adam Morgan, Michael Clark, Bryan Hart, Sasha Rohde, and Kevin Myers for assistance with molecular protocols. We are grateful to Rafael de Sá for providing us with advice regarding parsimony analysis. The manuscript was greatly improved by comments and suggestions from Heather Proctor and Andy Bohonak. This research was supported by grants from the Indiana Academy of Sciences, ARSAF, a faculty research award, and UExplore, an undergraduate research award at the University of Evansville.

References

Backeljau T., de Bruyn L., de Wolf H., Jordaens K., van Dongen S., Verhagen R. and Winnepenninckx B. 1995. Random amplified polymorphic DNA (RAPD) and parsimony methods. Cladistics 11: 119–130.

Böttger K. 1972. Vergleichende biologisch-ökologische Studien zum Entwicklungszyklus der Süsswassermilben (Hydrachnellae, Acari). II. Der Entwicklungzyklus von *Limnesia maculata* and *Unionicola crassipes*, International Revue Gesamten. Hydrobiologie 57: 263–319.

Dimock R.V. Jr. 1983. In defense of the harem: intraspecific aggression by male water mites (Acari: Unionicolidae). Ann. Entomol. Soc. Am. 76: 463–465.

Dimock R.V. Jr. 1985. Population dynamics of *Unionicola formosa* (Acari: Unionicolidae), a water mite with a harem. Am. Midl. Nat. 114: 168–179.

Dobson R. 1966. A survey of parasitic Unionicolidae (Arachnida: Acarina) of the Apalachicolan faunal region of the southern United States. Masters Thesis, Florida State University, Tallahassee, FL, 99 p.

Edwards D.D. and Dimock R.V. Jr. 1988. A comparison of the population dynamics of *Unionicola formosa* from two anodontine bivalves in a North Carolina farm pond. J. Elisha Mitch. Sci. Soc. 104: 90–98.

Edwards D.D. and Dimock R.V. Jr. 1997. Genetic differentiation between *Unionicola formosa* and *U. foili* (Acari: Unionicolidae): cryptic species of molluscan symbionts. Invertebr. Biol. 116: 124–133.

Edwards D.D. and Vidrine M.F. 1994. A new species in the water mite subgenus *Parasitatax* (Acari: Unionicolidae: *Unionicola*) from the North American freshwater mussel *Utterbackia imbecillis* (Bivalvia: Unionidae). J. Elisha Mitch. Sci. Soc. 110: 1–6.

Gledhill T. 1985. Water mites – predators and parasites. Freshw. Biol. Assoc. Ann. Rep. 53: 45–59.

Hadrys H., Balick M. and Schierwater B. 1992. Applications of random amplified polymorphic DNA (RAPD) in molecular ecology. Mol. Ecol. 1: 55–63.

Hevers J. 1978. Interspezifische Beziehungen zwischen *Unionicola*-Larven (Hydrachnellae, Acari) und Chironomidae (Diptera, Insecta). Verhandlungen Gesellschraft für Oekologie 7: 211–217.

Hevers J. 1980. Biologisch-ökologische Untersuchungen zum Entwicklungszklus der in Deutschland auftretenden *Unionicola*-Arten (Hydrachnellae, Acari). Arch. Hydrobiol. Suppl. 57: 324–373.

Jones R.K.H. 1978. Parasitism by *Unionicola* spp. larvae on chironomids. Hydrobiologia 60: 81–87.

LaRochelle P.B. and Dimock R.V. Jr. 1981. Behavioral aspects of host recognition by the symbiotic water mite *Unionicola formosa* (Acarina: Unionicolidae). Oecologia 48: 257–259.

Mitchell R.D. 1955. Anatomy, life history, and evolution of the mites parasitizing freshwater mussels. Miscellaneous Publications Museum of Zoology, University of Michigan 89: 1–28.

Ross K.G. 2001. Molecular ecology of social behavior: analyses of breeding systems and genetic structure. Mol. Ecol. 10: 265–284.

Smith I.M. and Oliver D.R. 1986. Review of parasitic associations of larval water mites (Acari: Parasitengona: Hydrachnida) with insect hosts. Can. Entomol. 118: 407–472.

Swofford D.L. 2002. PAUP*4.0b10: Phylogenetic Analysis using Parsimony (* and other Methods). Sinauer Associates, Sunderland, MA.

Vidrine M.F. 1996. North American *Najadicola* and *Unionicola*: Diagnoses and Distributions. Gail Q. Vidrine Collectibles, Eunice, Louisiana, 356 p.

Williams J.G., Hanafey M.K., Rafalsky J.A. and Tingey S.V. 1993. Genetic analysis using random amplified polymorphic DNA markers. Meth. Enzymol. 51: 704–737.

Experimental and Applied Acarology **34:** 79–93, 2004.

Recapture of male and female dragonflies in relation to parasitism by mites, time of season, wing length and wing cell symmetry

MARK R. FORBES[1,*], KATHERINE E. MUMA[2] and BRUCE P. SMITH[2]

[1]*Department of Biology, Carleton University, 1125 Colonel By Drive, Ottawa, Ontario, Canada KlS 5B6;* [2]*Biology Department, Ithaca College, 953 Danby Road, Ithaca, NY 14850-7278, USA;* **Author for correspondence (e-mail: mforbes@ccs.carleton.ca; phone: +1-613-520-2600 ext. 3873; fax: +1-613-520-3539)*

Key words: Arrenuridae, Dragonfly, Libellulidae, Mark-recapture, Parasitism, Survival, Symmetry

Abstract. For aquatic mites parasitic on dragonflies, completion of their life cycle depends on their being returned to appropriate water bodies by their hosts, after completion of engorgement. We examined whether differences among hosts in timing of emergence or phenotypic attributes might affect their probability of return to an emergence pond, and hence success of mites. Parasitized males and females of the dragonfly *Sympetrum obtrusum* (Hagen) did not differ in overall recapture rates. Females that had wing cell symmetry and emerged early were more likely to be recaptured than females that emerged later or had wing cell asymmetry, but there were no consistent relations between these variables and parasitism by mites. No such relations between wing cell asymmetry, emergence date, and recapture likelihood were found for males. Using randomization tests, we found that mean intensities of *Arrenurus planus* (Marshall) mites at host emergence were the same for recaptured females and females not recaptured; however, males that were recaptured had lower mean intensities of mites at emergence than males not recaptured. Further, mature females carried more mites than mature males, and the latter had fewer mites than newly emerged males not recaptured. Biases in detachment of engorging mites do not explain the differences in parasitism between mature males and females, nor the differences in mite numbers between mature males and newly emerged males that were not recaptured. Rather, heavily parasitized males appear to disperse or die and are not recaptured, which should have implications for dispersal of mites and fitness of male hosts.

Introduction

For dragonflies and damselflies, various factors have been implicated in affecting fitness or its supposed correlates such as lifespan, lifetime fecundity, lifetime mating success, shorter-term fecundity or mating success, dispersal, residency at territories, and recapture or return rates (reviewed by Corbet 1999). These factors include body size (Fincke 1982; Banks and Thompson 1987; Anholt 1990; Sokolovska et al. 2000; but see Thompson and Fincke 2002), fluctuating asymmetry (Harvey and Walsh 1993; Córdoba-Aguilar 1995; but see Forbes et al. 1997 and Carchini et al. 2000), and heterozygosity (Carchini et al. 2001). In addition, emergence date (Cordero Rivera 2000), and weather (Banks and Thompson 1987; Thompson 1990) have been shown to relate to surrogates of fitness in odonates.

Several studies have shown that such putative determinants of fitness can be interdependent. For example, in one species of damselfly, heterozygosity relates to body size, but only the former appears a significant predictor of short term mating success when variables are factored in relation to one another (Carchini et al. 2001). In another damselfly, wing cell asymmetry increases seasonally (Hardersen et al. 1999) and in many temperate damselflies, body size declines seasonally (e.g., Corbet 1999, and references therein). Such interrelationships make it difficult, using observational studies, to infer the independent contributions of these factors to fitness of insects being studied.

Of particular interest is how parasitism by water mites (Acari: Hydrachnida) influences host fitness because dragonfly and damselfly hosts have to return to appropriate water bodies to reproduce, following their pre-reproductive periods (Corbet 1999). Water mites also have to return to appropriate water bodies to complete their life cycle (Smith 1988). Parasitism by mites, however, is not independent of other factors. The number of mites on male and female dragonflies is related to factors such as emergence date (Forbes and Baker 1991; Forbes et al. 1999), time of season (Conrad et al. 2002), host age (Smith and Cook 1991), body size at emergence (Forbes and Baker 1990), nutritional deprivation during the larval stage (Leung et al. 2001), and fluctuating asymmetry (Bonn et al. 1996). Such co-variation makes it difficult to disentangle the influence of mites from the influence of host phenotypic characters, or other factors, on host fitness.

Despite these difficulties, several studies have shown that mites are associated with reduced survival of some damselflies during simulated bouts of inclement weather (Leung and Forbes 1997; Braune and Rolff 2001, but see Forbes and Leung 1995). Mites are also associated with reduced flight ability of hosts (Reinhardt 1996; suggested also by Conrad et al. 2002), and altered flight behaviour (Rehfeldt 1995, cited in Corbet 1999). In some host/parasite relationships, mites are correlated with reduced short-term mating success (Forbes 1991; Rehfeldt 1995; Andrés and Cordero Rivera 1998, cited in Corbet 1999; but see Rolff et al. 2000) and lowered fecundity (e.g., Forbes and Baker 1991). Thus, mites appear to impact their dragonfly hosts negatively, as expected from their nutritional requirements, and as evidenced by mite destruction of host tissue (Åbro 1979, 1982). However, the effects are not consistent.

The lack of consistent findings between mite parasitism and various measures of fitness of dragonfly hosts is not surprising. These associations *do* share certain aspects of natural history. For example, most involve species of *Arrenurus* that attach to larval dragonflies phoretically, but do not start parasitizing until the larval host ecloses to an adult (Smith 1988; Corbet 1999). However, these associations are vastly different even within the subgenus *Arrenurus* in terms of prevalence and intensity of infections (e.g., Forbes and Baker 1991; Bonn et al. 1996; Andrés and Cordero Rivera 1998; Yourth et al. 2001; Conrad et al. 2002; M.R. Forbes, personal observations, 1986–1994). Further, the degree to which individual mites engorge on their odonate hosts varies considerably within this subgenus, at a locality in Eastern North America (5–140×

their initial volume depending on the mite species; J. Moran and B.P. Smith, unpublished data). There also can be confusion as to the identification of mite species: there may be 20 or more species of *Arrenurus* in a water body and multiple interspecific parasitism of odonate hosts is common (B.P. Smith, personal observations, 1987–2002). This latter problem makes it difficult to ascribe changes in host behaviour, reproduction or host survival to individual mite species.

Other problems beset researchers interested in fitness effects of parasitic water mites on dragonfly hosts. In many associations, doing experimental infections is impractical. As such, researchers typically rely on natural infestations to investigate effects of mites on host survival or reproduction or host responses to mites (but see Léonard et al. 1999; Rolff 1999). Further, many researchers were unable to distinguish uninfected mature hosts from ones that were infected, but that lost their mites due to detachment (e.g., Robinson 1983; but see Forbes 1991 where accurate scoring of previous parasitism is possible). In cases where accurate scoring is not possible, researchers treating previously parasitized individuals as 'unparasitized' or 'lightly parasitized' may come to erroneous conclusions about the importance of mites to reproduction, survival and behaviour of their hosts. One way around this problem is to uniquely mark hosts as tenerals and to record variables of interest in relation to degree of mite parasitism initially experienced by individuals.

In this study, we assess whether likelihood of recapture of male and female *Sympetrum obtrusum* (Hagen) dragonflies, marked as tenerals, was dependent on their numbers of *Arrenurus planus* Marshall mites, wing cell asymmetry, emergence date, and/or wing length. We also explored the interrelations among these variables. We assessed whether numbers of mites differed between mature females and males, and explored whether any such differences existed for teneral dragonflies. It is important to look for relationships among these variables and the fate of teneral dragonflies with differing degrees of parasitism. This information might indicate whether individuals that were heavily parasitized at emergence were more likely to disperse (cf. Conrad et al. 2002) or die (cf. Braune and Rolff 2001) compared to individuals that were lightly parasitized at emergence.

We note that this comparison of numbers of mites on teneral versus mature males and females occurs after life history events that might influence initial degree of parasitism. Those events include frequency of encounters between mites and host larvae, grooming by host larvae to remove mites, or differences in transfer success of mites on to the newly eclosing host. It is important to realize that it is still not possible, with such studies, to disentangle the effect of initial condition of the host from subsequent parasitism (see Forbes and Baker 1990; Leung et al. 2001), and thus its potential influence on host survival or reproduction. Notwithstanding, it is still possible to explore the fate of heavily parasitized individuals and this information is of paramount importance to understanding constraints on mite success.

The host/parasite association chosen for our study is well suited for investigations into the determinants of the success of mites. First, *A. planus* mites can frequently number over 100 mites on single dragonfly hosts (see results) and mites engorge ca. 140× their initial volume (Forbes et al. 1999). Although scars left by mites are not as well demarcated as in other mite–dragonfly associations (Forbes 1991), one can still determine whether individuals have started losing their mites. We restricted samples of mature adults to those that had not yet shown signs of mite loss. Finally, *A. planus* show evidence of genetic differentiation between localities, suggesting that post-glacial dispersal has been limited as compared to other mite–insect associations (Bohonak 1999). This information is important from the viewpoint of local adaptation to hosts, which is especially relevant to parasites such as *A. planus* which have a broad host range (see Lajeunesse and Forbes 2002).

Materials and methods

Our study site was Yezerinac's Pond, a small ephemeral pond in Eastern Ontario Canada near the Queen's University Biological Station (QUBS), which is described elsewhere (Forbes et al. 1999, 2002). Several dragonfly and damselfly species emerge from this pond including *Sympetrum* dragonflies (Libellulidae) and *Lestes* damselflies (Lestidae). (Forbes et al. 1999; Yourth et al. 2001). This pond also supports *A. planus* (Acari: Arrenuridae) mites, but no other mite species known to parasitize odonates. These mites are generalists in that they attach to *Sympetrum* spp. dragonflies and to a lesser extent to *Lestes* spp. damselflies at our study site (Forbes et al. 2002). The mites are also ephemeral pond specialists, which has implications for their natural history (Wiggins et al. 1980). In particular, their large engorgement size enables these mites to diapause for many months as protonymphs, and to minimize deutonymphal feeding.

Similar to other members of the superfamilies Arrenuroidea, Hygrobatoidea, Lebertioidea, larval *A. planus* first attach to host larvae and are phoretic on them (Smith 1988, 1999). Once larval hosts emerge from the water and start to eclose into adults, mites crawl onto the newly formed imago. Mites pierce the host's cuticle using their chelicerae and secrete a stylostome or feeding tube (Smith 1988). In other odonate–mite associations, hosts may neutralize the stylostome(s) through melanotic encapsulation (Åbro 1982; Yourth et al. 2002; but see Forbes et al. 1999). If no or insufficient resistance is mounted, the mite engorges and later drops off when the host returns to the water for reproduction (e.g., Rolff and Martens 1997; Forbes et al. 1999).

Like all other 'derivative' water mites, these mites start to parasitize at host emergence, and engorgement is synchronous although detachment can occur over several days (Smith 1999). We have confirmed that mite detachment can occur over several days in this species with mark and recapture of uniquely marked individuals. Those mature individuals which have discolouration on the undersurface of their thorax or where mites were detaching at capture, could not be scored accurately for previous mite numbers. These were not included in our counts of mature hosts with engorging or engorged mites.

We netted dragonflies from June 7 to August 9, 1999. For each dragonfly netted, we counted mites on the thorax and abdomen using a 10× or 20× loupe. The only other species at the site that could be confused with *S. obtrusum* was *S. internum* Montgomery. We identified host species using facial colouration, male secondary genitalia, and female vulvar laminae and aged them as teneral or mature adults using criteria provided by Walker and Corbet (1975). We were confident that the dragonflies we scored as tenerals emerged from our study pond and were less than 2 days old when first netted. They were invariably weak flyers whose wings and bodies had not yet dried and hardened. As mentioned elsewhere (Forbes et al. 2002), the mites on teneral dragonflies were not engorged which is typical for mites that have just made the transition from host larva to recently eclosed adult (e.g., Mitchell 1968).

We used 259 tenerals (188 females and 71 males) to compare number of mites on newly emerged males and females. We scored emergence date for each teneral dragonfly (June 7, 1999 was scored as day 1). For each dragonfly, we also measured wing length by holding the right forewing onto a piece of paper and marking positions of the tip and nodus and measuring this distance using digital calipers (± 0.1 mm). For wing cell asymmetry measures, the number of cells from the nodus to the pterostigmata on the forward margins of the left and right forewings was counted. Difference in these measures gave an estimate of wing cell asymmetry (see also Hardersen et al. 1999; Yourth et al. 2002).

Tenerals and mature adults were uniquely marked. We used spots of blue, red, yellow, orange or purple acrylic paint on their left and/or right hind and/or fore wings. The marking protocols were such that each dragonfly received no more than two dots per wing; position of dots on wings corresponded to numbers (1, 2, 4, 8, 16, 32, 64, 128) which were summed to give an identity number. For example, a dot at the 64, 8, 4, 2 and 1 position referred to number 79. For each colour code, we could mark 255 dragonflies. We also used sex of host as a category, which provided us with many more unique combinations than needed.

We captured an additional 95 males and 65 females with mature colouration (following descriptions in Walker and Corbet 1975) that were parasitized with mites which were typically engorging or fully engorged where the mites scored 3–5, as in Forbes et al. (1999). Widths of fully engorged mites (score 5) ranged from 0.88 to 0.91 mm ($n = 7$, Forbes et al. 1999). We excluded mature adults where there were obvious signs of mite loss, either appearing as discolouration of thoracic or abdominal areas indicating scarring (cf. Forbes 1991), or individuals where mites detached during capture. Including this restricted sample of *S. obtrusum* males and females, that had mature colouration and their initial mite burdens, allowed us to compare mite numbers on a larger sample of dragonflies than we could have obtained solely through recapture of dragonflies we had marked when they were teneral. These mature dragonflies were either returning to Yezerinac's Pond following emergence without having been marked, or immigrating from elsewhere (see Michiels and Dhondt 1991 for work on *S. danae* where many individuals were expected to be immigrants).

Two main types of comparisons could be made: longitudinal analyses on dragonflies marked as tenerals in relation to whether or not they were subsequently recaptured, and comparisons between cohorts. For comparisons between males and females in probability of their recapture and for comparisons of wing cell asymmetry in relation to recapture for males and females separately, chi-square analysis were done. Student *t*-tests were used to evaluate potential relationships between timing of recapture in relation to sex of host, and for both wing length and emergence dates in relation to recapture, separately by sex of host. For data on recapture in relation to numbers of mites for teneral males and females, and on differences between numbers of mites for mature males and females, randomization tests were used. All analyses were conducted using JMP (SAS institute 1995) except for the randomization tests, which were done in visual basic. For the randomization tests, we tested whether mean mite numbers among the six categories of hosts differed from random expectation. The categories included males and females marked as tenerals and recaptured versus not, and males and females marked as mature adults upon first capture. This analysis is a distribution free method that allowed us to compare all groups simultaneously, despite groups having frequencies of mite numbers that did not conform to assumptions of normality (see results).

Results

All tenerals marked and recaptured were parasitized. Of 188 teneral females marked uniquely, 15 were recaptured (ca. 8%), whereas nine (or 12.7%) of 71 teneral males were recaptured. There was no difference between the sexes in the likelihood of recapture ($X^2 = 1.35$, $df = 1$, $p = 0.25$). However, the average interval between marking and first recapture of females (2.2 days ± 1.6 days) differed from that of males (10.1 ± 2.1 days, unpaired $t = -2.94$, $df = 22$, $p < 0.01$). Thus, although females marked as tenerals were equally likely to be recaptured as were males, males were first recaptured at longer intervals following their marking.

Of the 71 males marked as tenerals, four were not parasitized. The prevalence of parasitism by *A. planus* mites for teneral males was 94.4% (95% confidence limits calculated following Zar 1996: 86.2–98.4%). For teneral females, the result was consistent with nine of 188 individuals not having any mites: the prevalence of parasitism for females was 95.2% (91.1–97.8%).

There was an interesting relationship between the recapture of tenerals and the degree of parasitism. Abundance of mites on teneral males and females, regardless of whether or not they were recaptured, were non-normal (Shapiro-Wilk's values ranged from 0.72 to 0.82; *p*-values < 0.01, Figure 1). We thus completed a randomization test on the *F* value (from a one way ANOVA with

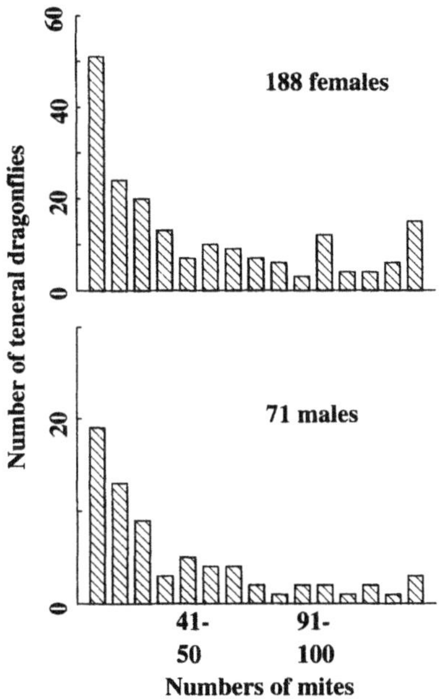

Figure 1. Frequency of numbers of *Arrenurus planus* mites found on 188 teneral female and 71 teneral male *Sympetrum obtrusum* dragonflies. Non-normal distributions (see text for details) were also found for mite numbers on mature females and males (data not shown). Mite numbers are organized into the following consecutive bins (0–10, 11–20, 21–30, 31–40, 41–50, 51–60, 61–70, 71–80, 81–90, 91–100, 101–125, 125–150, 151–175, 176–200, and 200+ mites). Only bins 41–50 and 91–100 are labelled.

group as a factor) to determine if there were any differences between the groups for mean mite numbers (Manly 1997). There was a significant difference between the groups (5000 Permutations $F = 5.46$, $p = 0.0006$; Figure 2). To determine which groups were different, we again used a randomization test on the Tukey's q statistic (Zar 1984). Teneral males and females, recaptured teneral females, and mature females with no evidence of mite loss, did not differ significantly in mite numbers (5000 Permutations, $p > 0.05$). However, these individuals had significantly more mites than the recaptured males and mature males ($p = 0.0001$). The median intensities of parasitism by mites for each category of host (and interquartile ranges) support these general findings (Table 1).

Next we investigated differences between dragonflies that were recaptured versus not, and whether it might account for spurious relationships between recapture likelihood and degree of parasitism by mites. Again, we considered males and females separately. Thirteen (or ca. 87%) of 15 recaptured females had wing cell symmetry. That is, they showed no difference in cell counts of left and right forewings. In comparison, 108 (or ca. 62%) of 173 females not recaptured were symmetrical ($X^2 = 4.1$, $df = 1$, $p < 0.05$). For males, 37 (or

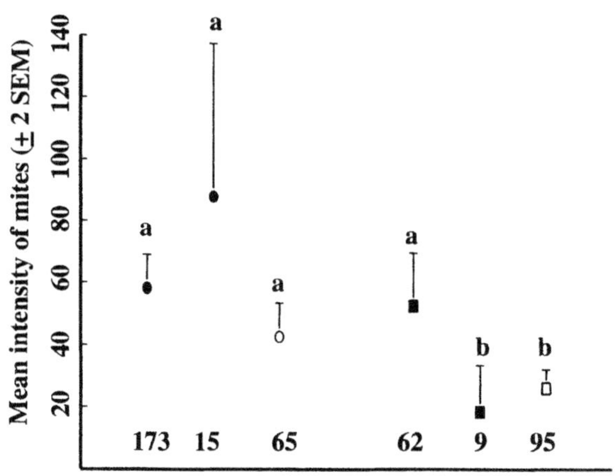

Figure 2. Mean intensity of *Arrenurus planus* mites observed on *Sympetrum obtrusum* individuals marked as tenerals (filled symbols), or individuals initially marked as mature adults (open symbols). In this figure, circles refer to female, and squares refer to male, *S. obtrusum*. These means are then compared to means from a randomization test to see which categories of hosts have fewer or more mites than expected (±2 SEM are presented for illustrative purposes only since these errors are not used in formulation of test statistics). Sample sizes, immediately above the *x*-axis, can be used to identify each category of host (cf. Table 1). There were 173 females marked as tenerals that were not recaptured; and 15 additional females marked as tenerals that were recaptured. There were 62 males marked as tenerals that were not recaptured; and nine additional males marked as tenerals that were recaptured. There were 65 and 95 females and males that were first captured and marked as individuals with mature colouration. Means that are not significantly different from one another based on the randomization test have the same letter designation (above the error bars). See text for details of analysis and see Table 1 for non-parametric descriptions of median mite intensities, for each category of host.

Table 1. Summary statistics of parasitism by *Arrenurus planus* mites for recaptured females and females not recaptured (F), and for recaptured males versus males not recaptured (M), after all were marked uniquely as tenerals. Same summary statistics are also presented for individuals first marked as mature adults.

Comparison		Status	*N*	Median (IQR)	Full range
Tenerals	F	Recaptured	15	36.0 (5.0–108.0)	2–379
		Not recaptured	173	28.0 (10.0–77.5)	0–382
	M	Recaptured	9	11.0 (4.0–28.0)	0–344
		Not recaptured	62	27.0 (10.5–62.5)	4–73
Mature adults	F	Marked as matures	65	30.0 (8.0–69.0)	1–255
	M	Marked as matures	95	14.0 (3.0–41.0)	1–133

Comparisons of mite numbers are also made between mature females and mature males, for which no mite loss had occurred. Medians, interquartile ranges (IQR) and full ranges of mite numbers are reported. *N* refers to the number of hosts.

ca. 60%) of 62 males not recaptured had wing cell symmetry compared with a similar proportion of 6 (or 67%) of 9 recaptured males ($X^2 = 0.16$, $df = 1$, $p = 0.69$). Thus, females with wing cell asymmetry were less likely to be recaptured as compared to symmetrical females (asymmetries in wing cell counts ranged from 1 to 5). This result did not hold for males.

We next examined recapture of tenerals in relation to emergence date and wing length since both body size and emergence date has been related to degree of mite parasitism in other studies (Forbes and Baker 1991). For females, there was a significant difference in emergence dates of those recaptured versus not. The 15 recaptured females emerged earlier on average than females not recaptured (Table 2). For males, there was no significant difference in emergence dates of those recaptured versus not. (Table 2). Although not statistically significant, recaptured females tended to have longer wings than females not recaptured (Table 2). This relationship was not seen in males (Table 2).

Finally, we examined how wing cell asymmetry, date of emergence, and size related to one another and to recapture likelihood. Emergence dates of 121 females with wing cell symmetry were compared with emergence dates of 67 females that showed some degree of wing cell asymmetry, using logistic regression. The same comparison was done for 48 males with wing cell symmetry and 28 males showing some degree of wing cell asymmetry (again, wing cell asymmetries ranging from 1 to 5). The regression coefficients for Julian date for both females and males were not significant (females: coefficient = −0.008, $df = 1$, $X^2 = 0.15$, $p = 0.69$ and males: coefficient = −0.04, $X^2 = 2.35$, $p = 0.13$). Thus, emergence date had no relation to symmetry for either males or females.

In comparison, wing length declined with emergence date for females (coefficient = −0.010, $t = -2.2$, $p < 0.05$) but not for males (coefficient = −0.005, $t = -0.76$, $p = 0.45$). Further, mite numbers declined significantly and near significantly with emergence date for females and males, respectively (Spearman $r = -0.252$, $p < 0.001$ and Spearman $r = -0.209$, $p = 0.067$). Dragonflies with fewer mites consistently emerged later in the season.

Table 2. Comparison between 15 recaptured females and 173 females not recaptured, and between nine recaptured males versus 62 males not recaptured, after all were marked uniquely as tenerals.

Variable		Status	Values	Test stat. (p-value)
Emergence date	F	recaptured	11.2 ± 1.9[a]	$t_{186} = 3.22$ (0.005)
		not recaptured	17.4 ± 0.54	
	M	recaptured	16.7 ± 2.9	$t_{69} = 0.87$ (0.38)
		not recaptured	19.5 ± 1.1	
Wing length	F	recaptured	12.63 ± 0.11[b]	$t_{186} = -1.742$ (0.080)
		not recaptured	12.4 ± 0.036	
	M	recaptured	12.5 ± 0.17	$t_{69} = 0.22$ (0.82)
		not recaptured	12.5 ± 0.065	

Comparisons are made with respect to emergence date and wing length (in mm) using t-tests separately by sex of host.

[a] Mean date of emergence presented ± 1SE.

[b] Mean wing length presented ± 1SE.

Discussion

Exploring whether naturally occurring parasites have effects on fitness of their hosts is a daunting task. There is the problem, seen in many studies, that initial health or condition of hosts may influence degree of subsequent parasitism and that it may be this underlying 'condition' rather than parasitism *per se* that influences host fitness. Many other factors including time of season might influence the degree to which parasites are seen to exert their effects. Alternatively, time of season may have a direct effect on host fitness and degree of parasitism is only a correlate of host fitness, because of the direct relation between degree of parasitism and time of season.

Despite these difficulties, it is nonetheless important to consider measures of fitness for hosts differing in naturally occurring levels of parasitism and whether these differences are also related to differences among hosts in either phenotypic attributes or timing of emergence. In brief, such studies can help researchers determine the conditions affecting mite success. For example, if heavily parasitized hosts or smaller hosts are less likely to be recaptured, then mites on those hosts may not achieve high relative success.

The current study is framed in the context of the factors that might affect both degree of parasitism and host fitness. Our results show that various factors are important in determining recapture of *S. obtrusum* at an emergence pond, only some factors being interdependent. Further, the importance of such factors is dependent on sex of host. Our most salient finding was that males that were recaptured tended to have fewer mites than males also marked as tenerals, but not recaptured. Such relationships were not found for females. If recaptured males and females were representative of mature dragonflies returning to the ponds then we might expect a sex bias in parasitism of mature adults by mites, which is what was observed. Mature males had significantly fewer mites than mature females. We also observed that teneral males not recaptured had more mites than mature males: an age bias not seen in females. These results suggest that heavily parasitized males in this species were either more likely to disperse or to die than were lightly parasitized males.

We do not believe that males were more likely to lose their mites, since only mature individuals where no mite loss was suspected were included in these analyses. Moreover, we could verify that earlier mite counts were accurate and, for individuals marked as tenerals and subsequently recaptured. Others have suggested mite losses for female hosts because of 'wetting' mites during oviposition (Mitchell 1968; McKee et al. 2003) Clearly, this pattern was not observed. It is also important to note that *S. obtrusum* females have been observed ovipositing and losing mites while flying above damp soil, without wetting their abdomens (M.R. Forbes, personal observations, 2000). We do not expect biases in detachment of *A. planus* mites for male or female *S. obtrusum*, as has been suggested for other mite–dragonfly associations.

Our main supposition is that heavily parasitized males either show greater dispersal or reduced survival. In S. *obtrusum*, most *A. planus* mites attach on the venter of the thorax. In fact, abdominal mites are typically found only on the most heavily parasitized individuals that have thoraces extremely crowded with mites (≥50 mites, K.E. Muma unpublished data). The feeding tubes of these mites would be expected to be associated with digestion of tissue either near, or in, the flight musculature, as noted by others (reviewed in Corbet 1999). Additionally, one might expect such heavy burdens to impose a load cost on flying odonates. The more parsimonious explanation is that survivorship of heavily parasitized males is reduced and this represents a cost for both the host and the mites. However, Conrad et al. (2002) showed a relationship in which parasitized males of a damselfly were more likely to be recaptured away from emergence ponds than their unparasitized counterparts (but see Smith and McIver 1984 for the reverse pattern in mosquitoes). Whether or not mites influence dispersal of their hosts needs to be investigated more fully.

Another salient finding was that females were more likely to be recaptured if they emerged early in the season. This relationship could explain why recaptured females also tended to have longer wings, but not why they were also more symmetrical, since emergence dates were only correlated (inversely) with wing lengths for females (cf. Hardersen et al. 1999). Studies have had mixed conclusions as to whether wing symmetry relates to indices of fitness in odonates (see introduction). Like most studies on asymmetry, natural variation was used, and, in this specific instance, symmetry scores were based on cell counts rather than wing length fluctuating asymmetry. Future mark-recapture studies in insects should continue to address this problem for two reasons: first, such symmetry scores are easy to obtain and highly repeatable. Second, these scores are obtained for a functional trait that is often sexually dimorphic in size and/or shape and that is involved in flight-based return or dispersal. At present, we do not consider wing cell asymmetry further since it was unrelated to emergence date in either sex: emergence date was, however, related inversely to mite numbers.

It is possible that the relationship between likelihood of recapture and emergence date for females is an artifact. Those dragonflies marked early in the season have more time to be recaptured than those marked later in the season. However, if this relationship was strong, then there should have been a correlation for males, but there was not. It is likely that other studies will report relationships between recapture likelihood and emergence date simply because of weather-related variation in survival or activity (cf. Conrad et al. 2002). We would expect both positive and inverse correlations to be found, e.g., dragonflies emerging early but subjected to inclement weather would have reduced representation in recapture totals in some seasons. Following from this line of reasoning, variability among cohorts in exposure to inclement weather could obscure any links between wing length or asymmetry and recapture or survival.

It is important that females emerging later with fewer mites may actually be less represented in recaptures in at least some years. If this result were repeatable then it would add credence to the idea that later emerging individuals of insects from temperate climates have lowered fitness (they tend to be smaller at emergence and may mature after the peak of the flight season). We currently do not know the extent to which host fitness (i.e. host ability to survive or to frequent appropriate water bodies for reproduction) is affected by time of emergence. Our results do suggest, however, that the consequences, for fitness of mites may be worth exploring further. There also have been suggestions that mites active later in the season may suffer greater costs and have low fitness, simply because they are less able to find hosts (cf. Rolff et al. 2001) or because they encounter greater resistance from hosts (Yourth et al. 2002).

It is important to consider why such costs, if they occur, are not countered by adaptations of the mite to avoid later emerging hosts. As mentioned elsewhere (Forbes et al. 1999) the likelihood of host discovery is probably quite low. Also, the costs to host discrimination at this stage may entail risks of failing to attend suitable hosts. Related to this point, *A. planus* mites are frequently found on *S. internum*, yet this host is almost entirely resistant and mites fail to engorge (Forbes et al. 2002). Thus, mites seem unable to avoid unsuitable hosts at the host species level. It is also interesting that for *S. internum*, there are no biases in parasitism for mature male and female dragonflies as was seen for *S. obtrusum*. We suspect that the costs of parasitism are greater for *S. obtrusum* than *S. internum*, and that the costs of parasitism may fall disproportionately on males of *S. obtrusum* (cf. Braune and Rolff 2001).

Taken together, we argue for mite-induced reduced survivorship of male dragonflies. The possibility still exists that such males have greater nutritional requirements and forage further from emergence ponds (cf. Conrad et al. 2002) or that such males are simply targets of greater intraspecific aggression and are pushed farther away (Rehfeldt 1995, cited in Corbet 1999). These alternatives cannot yet be dismissed. At any rate, mites on heavily parasitized hosts of at least one sex appear to suffer reduced fitness either through reduced likelihood of returning to their natal pond or finding a suitable alternative pond. We also suggest that mites active later in the season may suffer reduced fitness based on previous studies. This study adds another possible consequence of being active later in the season, i.e. attending to hosts that are less likely to return to ponds. Taken collectively, such results would imply strong selection for early breeding activity and egg-laying of *A. planus*, a supposition we are currently investigating.

Acknowledgements

We thank Frank Phelan and Floyd Connor at the Queen's University Biological Station for logistical support and access to the Hilda S. Pangman Conservation Reserve. The project was supported by a Summer Faculty Research Grant from the Provost at Ithaca College to KEM. Christopher Yourth and Jennifer Moran assisted in the field. Tonia Robb performed the randomization tests.

References

Åbro A. 1979. Attachment and feeding devices of water mite larvae (*Arrenurus* spp.) parasitic on damselflies (Odonata, Zygoptera). Zool. Scripta, 8: 221–234.

Åbro A. 1982. The effects of parasitic water mite larvae (*Arrenurus* spp.) on zygopteran imagoes. J. Inv. Pathol. 39: 373–381.

Andrés J.A. and Cordero Rivera A. 1998. Effects of water mites on the damselfly *Ceriagrion tenellum*. Ecol. Entomol. 25: 103–109.

Anholt B.R. 1990. Size-biased dispersal prior to breeding in a damselfly. Oecologia 83: 385–287.

Banks M.J. and Thompson D.J. 1987. Lifetime reproductive success of females of the damselfly, *Coenagrion puella*. J. Anim. Ecol. 56: 815–832.

Bohonak A.J. 1999. Effect of insect-mediated dispersal on the genetic structure of postglacial water mite populations. Heredity 82: 451–461.

Bonn A., Gasse M., Rolff J. and Martens A. 1996. Increased fluctuating asymmetry in the damselfly *Coenagrion puella* is correlated with ectoparasitic water mites: implications for fluctuating asymmetry theory. Oecologia 108: 596–598.

Braune P. and Rolff J. 2001. Parasitism and survival in a damselfly: does host sex matter? Proc. R. Soc. Lond. B 268: 1133–1137.

Carchini G., Chirotti F., Domenico M. and Paganotti G. 2000. Fluctuating asymmetry, size and mating success in males of *Ischnura elegans* (Vander Linden) (Odonata: Coenagrionidae). Anim. Behav. 59: 177–182.

Carchini G., Chirotti F., Domenico M., Mattoccia M. and Paganotti G. 2001. Fluctuating asymmetry, mating success, body size and heterozygosity in *Coenagrion scitulum* (Rambur) (Odonata: Coenagrionidae). Anim. Behav. 61: 661–669.

Conrad K.F., Willson K.H., Whitfield K., Harvey I.F., Tomas C.J. and Sherratt T.M. 2002. Characteristics of dispersing *Ischnura elegans* and *Coenagrion puella* (Odonata): age, sex, size, morph and ectoparasitism. Ecography 25: 439–445.

Corbet P.S. 1999. Dragonflies: Behavior and Ecology of Odonata. Cornell University Press, Ithaca, New York, pp. 829.

Cordero Rivera A. 2000. An analysis of multivariate selection in a non-territorial damselfly (Odonata: Coenagrionidae). Etologia 8: 37–41.

Córdoba-Aguilar A. 1995. Fluctuating asymmetry in paired and unpaired damselfly males, *Ischnura denticollis* (Burmeister) (Odonata: Coengrionidae). J. Ethol. 13: 129–132.

Fincke O.M. 1982. Lifetime Mating Success in a Natural-Population of the Damselfly, *Enallagma hageni* (Walsh) (Odonata, Coenagrionidae). Behav. Ecol. Sociobiol. 10: 293–302.

Forbes M.R. 1991. Ectoparasites and mating success of male *Enallagma ebrium* damselflies (Odonata: Coenagrionidae). Oikos 60: 336–342.

Forbes M.R. and Baker R.L. 1990. Susceptibility to parasitism: experiments with the damselfly *Enallagma ebrium* (Odonata: Coenagrionidae) and larval water mites, *Arrenurus* spp. (Acari: Arrenuridae). Oikos 58: 61–66.

Forbes M.R.L. and Baker R.L. 1991. Condition and fecundity of the damselfly, *Enallagma ebrium* (Hagen): the importance of ectoparasites. Oecologia 86: 335–341.

Forbes M.R. and Leung B. 1995. Pre-fabricated dining shelters as outdoor insectaries, an assessment using *Enallagma ebrium* (Hagen) (Zygoptera: Coenagrionidae). Odonatologica 24: 461–466.

Forbes M.R., Leung B. and Schalk G. 1997. Fluctuating asymmetry in *Coenagrion resolutum* (Hagen) in relation to age and male pairing success (Zygoptera: Coenagrionidae). Odonatologica 26: 9–16.

Forbes M.R., Muma K.E. and Smith B.P. 1999. Parasitism of *Sympetrum* dragonflies by *Arrenurus planus* mites: maintenance of resistance particular to one species. Int. J. Parasitol. 29: 991–999.

Forbes M.R., Muma K.E. and Smith B.P. 2002. Diffuse coevolution: constraints on a generalist parasite favor use of a dead-end host. Ecography 25: 345–351.

Hardersen S., Wratten S.D. and Frampton C.M. 1999. Does carbaryl increase fluctuating asymmetry in damselflies under field conditions? A mesocosm experiment with *Xanthocnemis zealandica* (Odonata: Zygoptera) J. Appl. Ecol. 36: 534–543.

Harvey I.F. and Walsh K.J. 1993. Fluctuating asymmetry and lifetime mating success are correlated in males of the damselfly *Coenagrion puella* (Odonata: Coenagrionidae). Ecol. Ent. 18: 198–202.

Lajeunesse M.J. and Forbes M.R. 2002. Host range and local parasite adaptation. Proc. R. Soc. Lond. Ser. B 269: 703–710.

Léonard N.J., Forbes M.R. and Baker R.L. 1999. Effects of *Limnochares americana* (Hydrachnida: Limnocharidae) mites on life history traits and grooming behaviour of its damselfly host, *Enallagma ebrium* (Odonata: Coenagrionidae). Can. J. Zool. 77: 1615–1622.

Leung B. and Forbes M.R. 1997. Fluctuating asymmetry in relation to indices of quality and fitness in the damselfly, *Enallagma ebrium* (Hagen). Oecologia 110: 472–477.

Leung B., Forbes M.R. and Baker R.L. 2001. Nutritional stress and behavioural immunity of damselflies. Anim. Behav. 61: 1093–1099.

Manly B.F.J. 1997. Randomization, Bootstrap and Monte Carlo Methods in Biology, 2nd ed. Chapman & Hall, UK, 399 p.

McKee D., Harvey I., Thomas M. and Sherratt T.N. 2003. Mite infestation of *Xanthocnemis zealandica* (McLachlan) in a Christchurch Pond. New Zeal. J. Zool. 30: 17–20.

Michiels N.K. and Dhondt A.A. 1991. Characteristics of dispersal in sexually mature dragonflies. Ecol. Entomol. 16: 449–459.

Mitchell R. 1968. Site selection by larval water mites parasitic on the damselfly *Cercion hieroglyphicum* Brauer. Ecology 49: 40–47.

Rehfeldt G.E. 1995. Natürliche Feinde, Parasiten und Forpflanzen von Libellen. Wolfram Schmidt, Braunschweig.

Reinhardt K. 1996. Negative effects of *Arrenurus* water mites of the flight distance of the damselfly, *Nehalennia speciosa* (Odonata: Coenagrionidae). Aq. Insects 18: 233–240.

Robinson J.V. 1983. Effects of water mite parasitism on the demographics of an adult population of *Ischnura posita* (hagen) (Odonata: Coenagrionidae). Am. Midl. Nat. 109: 169–174.

Rolff J. 1999. Parasitism increases offspring size in a damselfly: experimental evidence for parasite-mediated maternal effects. Anim. Behav. 58: 1105–1108.

Rolff J., Antvogel H. and Schrimpf I. 2000. No correlation between ectoparasitism and male mating success in a damselfly: why parasite behavior matters. J. Insect Behav. 13: 563–571.

Rolff J. and Martens A. 1997. Completing the life cycle: detachment of an aquatic parasite (*Arrenurus cuspidator*, Hydrachnellae) from an aerial host (*Coenagrion puella*, Odonata). Can. J. Zool. 75: 655–659.

Rolff J., Vogel C. and Poethke H.J. 2001. Co-evolution between ectoparasites and their insect hosts: a simulation study of a damselfly-water mite interaction. Ecol. Entomol. 26: 638–645.

SAS Institute Inc. 1995. JMP User's Guide, Version 3. SAS Institute Inc., Cary, North Carolina.

Sokolovska N., Rowe L. and Johansson F. 2000. Fitness and body size in mature odonates. Ecol. Ent. 25: 239–248.

Smith B.P. 1988. Host–parasite interaction and impact of larval water mites on insects. Ann. Rev. Ent. 33: 487–507.

Smith B.P. 1999. Larval *Hydrachnidia* and their hosts: biological inference and population structure. In: Needham G.R., Mitchell R., Horn D.J. and Welbourn W.C. (eds), Acarology IX, Vol. 2, Symposia, Ohio Biological Survey. Columbus, pp. 139–144.

Smith B.P. and Cook W.J. 1991. Negative covariance between larval *Arrenurus* sp. and *Limnochares americana* (Acari: Hydrachnida) on male *Leucorrhinia frigida* (Odanata: Libellulidae) and its relationships to the host's age. Can. J. Zool. 69: 226–231.

Smith B.P. and McIver S.B. 1984. The impact of *Arrenurus danbyensis* Mullen (Acari: Prostigmata; Arrenuridae) on a population of *Coquillettidia perturbans* (Walker) (Diptera: Culicidae). Can. J. Zool. 62: 1121–1134.

Thompson D.J. 1990. The effects of survival and weather on lifetime egg production in a model damselfly. Ecol. Ent. 15: 455–462.

Thompson D.J. and Fincke O.M. 2002. Body size and fitness in Odonata: stabilizing selection and a meta-analysis too far? Ecol. Ent. 27: 378–384.

Walker E.M. and Corbet P.S. 1975. The Odonata of Canada and Alaska, Vol. 3. Univ. Toronto Press, Toronto, Ontario.

Wiggins G.B., MacKay R.J. and Smith I.M. 1980. Evolutionary and ecological strategies of animals in annual temporary ponds. Arch. Hydrobiol. Suppl. 76: 369–392.

Yourth C.P., Forbes M.R. and Smith B.P. 2001. On understanding variation in immune expression of the damselflies *Lestes* spp. Can. J. Zool, 79: 815–821.

Yourth C.P., Forbes M.R. and Smith B.P. 2002. Immune expression in a damselfly is related to time of season, not to fluctuating asymmetry or host size. Ecol. Ent. 27: 123–128.

Experimental and Applied Acarology **34:** 95–112, 2004.

Specificity of attachment sites of larval water mites (Hydrachnidia, Acari) on their insect hosts (Chironomidae, Diptera) – evidence from some stream-living species

PETER MARTIN
Christian-Albrechts-Universität zu Kiel, Zoologisches Institut, Olshausenstr. 40, D-24098 Kiel, Germany; (e-mail: pmartin@zoologie.uni-kiel.de; phone: +49-431/880-4136; fax: +49-431/880-4368)

Key words: Attachment site specificity, Host partitioning, Parasitism, Running water

Abstract. This study concerns the parasite–host associations of water mite larvae and their chironomid hosts in a small stream on the North German Plain. The different feeding sites on the host were tested as to whether they represented a strategy of the parasites regarding host partitioning. The attachment sites of nine ectoparasitic water mite species were observed in detail: *Aturus fontinalis*, *Atractides nodipalpis*, *Feltria rouxi*, *Hygrobates nigromaculatus*, *Protzia eximia*, *Sperchonopsis verrucosa*, *Sperchon clupeifer*, *S. setiger* and *Lebertia inaequalis*. *Aturus fontinalis*, *A. nodipalpis*, *F. rouxi* and *H. nigromaculatus* distinctly preferred sites on the abdomen of the host, whereas the other species preferred feeding sites on the thorax. The four species that preferentially attached to the abdomen of the host showed a distinct specificity for selected segmental and/or intersegmental regions. All species differed in their sites along the length of the abdomen. The order of attachment on the abdominal segments was, from anterior to posterior: *H. nigromaculatus*, *F. rouxi*, *A. fontinalis* and *A. nodipalpis*. The sites were analysed with regard to segmental and intersegmental attachment, the proportion of dorsal and ventral sites and the differences between attachment to different host species. Larvae attached to their hosts as single individuals showed only slight differences from the sites of mites on hosts that carried many mite larvae. The finding that less than 10% of the chironomids were parasitized by more than one water mite species suggested that, at least in the zoocoenosis of the studied collecting site, the interspecific competition for attachment sites was not strong. However, the specificity of attachment sites clearly had the potential of diminishing competition between water mite species by host partitioning. Intra- and interspecific competition for preferred sites and preparasitic constraints are discussed.

Introduction

Water mites (Hydrachnidia, Acari) have complex life cycles. Adults and deutonymphs are predators of insects, crustaceans, eggs and other mites (Proctor and Pritchard 1989). Most water mite species pass through an ectoparasitic larval stage on insect hosts of various orders (see Smith and Oliver 1986; Smith et al. 2001). Numerous larvae will often parasitize one host specimen and parasitism of a host individual by larvae of more than one species of water mite occurs regularly (Smith et al. 2001).

With regard to parasitism by water mite larvae, sympatry of a number of species is possible by the utilization of different host resources (see Lanciani 1970): larvae can either parasitize different host species or exploit the host in different biotopes, at different times or by attachment to different sites on the host. In the cases studied here, the host as a resource with different places for attachment of parasitic mites has been observed.

In terms of the numbers of species and individuals recorded as being parasitized by water mite larvae, the dipteran family Chironomidae is the most important host group of Hydrachnidia. Few extensive studies of water mites and their chironomid hosts are available: attachment sites are often vaguely given as being the tagmata of the host (Ullrich 1978) or mites are classified only at the genus level (LeSage and Harrison 1980). Detailed data on attachment sites are available almost exclusively for mite species from standing water, e.g. parasitizing Odonata, Hemiptera or Diptera (e.g. Booth and Learner 1978) and often focus on a single parasite and/or host species (Davids et al. 1977; Meyer 1985; Wiles 1987). The only detailed study of host-parasite associations in water mites from running waters is that of Efford (1963) who has found specific attachment sites of water mite species on their chironomid hosts.

The different distribution of attachment and feeding sites of larval water mites on their insect host has a phylogenetic component (Smith and Oliver 1986). However, preparasitic constraints and intraspecific and interspecific competition also play important roles in the observed patterns of attachment sites. When different larvae of a single water mite species really prefer distinct attachment sites, intraspecific competition takes place: in describing the behaviour of *Arrenurus* larvae parasitizing ceratopogonids (Diptera, Nematocera), Mitchell (1998) illustrates the intraspecific competition of these larvae during the ecdysis of the adult host from the pupa. Evidence for interspecific competition is provided by Efford (1963) who has observed the plasticity of attachment sites in cases of multiple parasitism on one single host. Nevertheless, the relevance of competition in determining placement on the host is not clear.

Here, data are presented regarding attachment sites based on a 1 year study of water mite parasitism in a small lowland stream where chironomids are the most important hosts for water mites (Martin 2000). Answers to the following questions have been sought. (1) Do different water mites exhibit a preference for selected attachment sites? (2) If so, is this preference more distinct when only a single mite larva attaches to a host? (3) Is there variability regarding these attachment sites when different hosts are compared? (4) Is the attachment to certain sites influenced by intra- and interspecific competition? (5) Are there other explanations for the observed patterns of attachment?

Material and methods

The data presented are based on collections taken during 1996 from a summer-warm lowland stream in Northern Germany (the Farver Au, 10°48′10″E; 54°15′58″N). A detailed description of the collecting site is given by Martin (1996, 1998). Emergence traps were used for sampling parasitized insects, mainly chironomids (Diptera, Nematocera, Chironomidae). The traps were emptied weekly from the beginning of May until the end of November 1996. Details of the design and information on the emergence traps are given in Martin (1998).

The attachment sites of the water mite larvae were noted before the larvae were detached and individually fixed for determination. The water mite larvae and their chironomid hosts were identified to species level. For the classification of mite larvae, descriptions of larval morphology from the literature and new results of rearing experiments were used (Martin 1998, 2000).

Attachment sites were categorized in two ways. First, they were broken into major tagmata (head thorax, abdomen). Second, attachment along the length of the abdomen was subdivided. For the statistical analysis, attachment site categories were set so that larvae attached intersegmentally between the thorax and abdomen and larvae attached to the first abdominal segment were summarized as ' ≤ 1', larvae attached intersegmental between the first and second abdominal segment and larvae attached to the second abdominal segment were summarized as ' ≤ 2', and so on.

For all statistical analyses, χ^2-tests of uniform distribution were applied by using STATISTICA 6.0, StatSoft Inc. The significance between observed and expected values (differences between dorsal and ventral attachment, non-homogeneous distribution of attachment sites) was calculated by the use of 2×2 tables and the χ^2-test with contingency tables. Bonferroni's correction was used to adjust for multiple testing (Sachs 1993).

Of the 5801 collected chironomids, 978 (= 18%) were found to be parasitized by water mite larvae. A detailed analysis of the parasite–host associations will be published elsewhere (Martin, in prep.) but some summary data are given here, considered necessary for a better understanding of the presented results. The summarized data for the most abundant water mite species and the three most abundant host taxa are as follows (for terminology see Margolis et al. 1982):

	Mean abundance (larvae/host)	Mean intensity (larvae/host)	Max. intensity (larvae/host)	Prevalence %
Aturus fontinalis	0.33	1.75	10	19.0
Atractides nodipalpis	0.21	1.49	7	13.8
Feltria rouxi	0.10	2.10	19	4.7
Hygrobates nigromaculatus	0.17	1.56	4	10.9

Results

In total, 1748 water mite larvae were analysed with regard to their exact attachment sites on their chironomid hosts. *Aturus fontinalis* Lundblad 1920, *Atractides nodipalpis* (Thor 1899), *Feltria rouxi* Walter 1907 and *Hygrobates nigromaculatus* Lebert 1879 were the most abundant species. Table 1 shows the attachment sites of nine species that were exclusively found on chironomids or, in the case of *Protzia eximia* (Protz 1896) and *Sperchon setiger* Thor, 1898, on chironomids and other host taxa (simuliid and psychodid dipterans, Trichoptera, see Martin 2000).

Two categories of chironomid parasites could be distinguished (Table 1): *P. eximia*, *Sperchonopsis verrucosa* (Protz 1896), *Sperchon clupeifer* Piersig 1896, *S. setiger* and *Lebertia inaequalis* (Koch 1837) were almost all attached to the thorax of the host. The coxa of the first legs was the most important attachment site. For several species, other sites could also occasionally be found (cervix: *P. eximia*, *S. clupeifer*, *S. setiger*, *L. inaequalis*; coxa of third leg: *P. eximia*, *L. inaequalis;* anterior segments of the abdomen: *P. eximia*, *S. clupeifer*). The other group of parasites was preferentially found on the abdomen of the host (both groups, χ^2-test: $p < 0.001$). This preference was shown by *A. fontinalis*, *A. nodipalpis*, *F. rouxi* and *H. nigromaculatus*. In this group, the rare attachment to thoracic segments was recorded only in one specimen of *A. fontinalis* and four of *H. nigromaculatus*.

Of the four species preferring attachment to the abdomen of the host, only a small number of specimens was found detached, i.e. larvae that had detached after the conservation of the parasitized hosts. In total, 221 detached larvae were recorded. The species preferentially attaching to the thorax of the host were found with a higher frequency detached than were the parasites of the abdomen of the host (comparison of all individuals of attached versus detached larvae: χ^2-test, $p < 0.001$). In the former, a maximum of 48.0% detached larvae was found (*S. verrucosa*), whereas in the latter, this value was 16.4% (*H. nigromaculatus*). In view of the distinctly higher number of individuals found parasitic on the abdominal segments, I will focus on these species in the following.

Considering all host specimens together, the four species with preference for abdominal attachments were predominantly attached ventrally (χ^2-test, $p < 0.001$), although with some variability from host species to host species: the larvae of *A. fontinalis* showed a preference for the dorsal abdomen of *Synorthocladius semivirens* (Kieffer 1909) but for the ventral abdomen of *Rheocricotopus fuscipes* (Kieffer 1909). Additionally, no significant preference for the ventral abdomen could be shown in *A. fontinalis* parasitizing *Paratrichocladius skirwithensis* (Edwards 1929) or other hosts. The same was observed in *A. nodipalpis* parasitizing *Polypedilum convictum* (Walker 1856) or other hosts. In *F. rouxi*, the preferred attachment to the ventral abdomen of the host was significant for all host species.

In the most abundant mite species, *A. fontinalis*, the preferred site was independent of the host sex (Table 2, no significance by χ^2). Neither the accumulation on dorsal attachment sites (*S. semivirens)* nor the accumulation on ventral sites (*R. fuscipes*) showed a bias based on host sex.

Differences were noted concerning the exact position of the abdominal attachment site on comparing the total number of mites (Table 1): *A. fontinalis*

Table 1. Occurrence of the chironomid parasitizing water mites at the various tagmata of the body of the host.

		Aturus fontinalis					*Atractides nodipalpis*					*Feltria rouxi*					*Hygrobates nigromaculatus*						*P. eximia*	*Sp. verrucosa*	*S. clupeifer*	*S. setiger*	*L. inaequalis*
		Synorthocladius semivirens	*Rheocricotopus fuscipes*	*Paratrichocladius skirwithensis*	Other hosts	Total	*Rheocricotopus fuscipes*	*Orthocladius* sp.	*Polypedilum convictum*	Other hosts	Total	*Paratrichocladius skirwithensis*	*Orthocladius* sp.	*Rheocricotopus fuscipes*	Other hosts	Total	*Micropsectra* spp.	*Tanytarsus* spp.	*Cladotanytarsus vandervulpi*	*Paratrichocladius skirwithensis*	Other hosts	Total	Total	Total	Total	Total	Total
Caput/thorax, dorsal (Cervix)																							6				
Caput/thorax, ventral																							3		1	1	1
Thorax	First leg					1																1	46	13	10	8	7
	Second leg																										
	Third leg																					3	5				1
																							***				
Abdomen	th/1		1			1						1				1	1					1					
Intersegmental	1/2	10	12	15	8	45		1			1	37	27	11	16	91	16	12	6	1	7	42					
	2/3	76	79	31	19	205	15			2	17	39	20	7	11	77	1	1	1			3					
	3/4	202	170	35	26	433	6			1	7	27	3	2	5	37	1			1	1	3					
	4/5	33	19	22	6	80	31	1	2	2	36	11	7			18											
	5/6	7	5	3		15	22	6	15	6	49	2	3			5											
	6/7				1	1	24	11	22	3	60	1			1	2											
	7/8						2	1	2	1	6																
		***	***	***	***		***	***	***	n.s.		***	***	*	**		***	**									
Abdomen	1				1	1	1				1		1		1	1	39	35	14	14	17	119	2				
Segmental	2						4				4			1	1	2	1	1	1	2	1	6			1		
	3						17			1	18		1	1	1	3											

Table 1. Continued

		Aturus fontinalis					*Atractides nodipalpis*					*Feltria rouxi*					*Hygrobates nigromaculatus*						*P. eximia*	*Sp. verrucosa*	*S. clupeifer*	*S. setiger*	*L. inaequalis*
		Synorthocladius semivirens	*Rheocricotopus fuscipes*	*Paratrichocladius skirwithensis*	Other hosts	Total	*Rheocricotopus fuscipes*	*Orthocladius* sp.	*Polypedilum convictum*	Other hosts	Total	*Paratrichocladius skirwithensis*	*Orthocladius* sp.	*Rheocricotopus fuscipes*	Other hosts	Total	*Micropsectra* spp.	*Tanytarsus* spp.	*Cladotanytarsus vandervulpi*	*Paratrichocladius skirwithensis*	Other hosts	Total	Total	Total	Total	Total	Total
Abdomen	4						26	3	1		30			1		2											
Segmental	5						87	6	1	3	97																
	6						57	14	4	1	76																
	7						21	12	5		38		1			1											
	8						1	1			2																
							***	*	n.s.								***	***	***	**	***						
Total attached larvae		328	287	106	61	782	314	56	52	20	443	118	63	23	36	240	60	51	22	18	27	178	62	13	12	9	9
Attached to abdomen (%)		100.0	99.7	100.0	100.0	99.9	100.0	100.0	100.0	100.0	99.8	100.0	100.0	100.0	100.0	100.0	98.3	96.1	100.0	100.0	96.3	97.8	3.2	0.0	8.3	0.0	0.0
Ind. abd. dorsally (%)		74.1	4.9	46.2	45.9	42.8	7.0	10.7	63.5	30.0	15.2	11.9	15.9	17.4	19.4	14.6	1.7	2.0	0.0	0.0	3.8	1.7					
Ind. abd. ventrally (%)		25.9	95.1	53.8	54.1	57.2	93.0	89.3	36.5	70.0	84.8	88.1	84.1	82.6	80.6	85.4	98.3	98.0	100.0	100.0	96.2	98.3					
		***	***	n.s.	n.s.	***	***	***	n.s.	n.s.	***	***	***	**	***	***	***	***	***	***	***	***					

For the first four species, the distribution of the larvae at the abdominal segments are shown for the most abundant host species. The results for all hosts were grouped together. The χ^2-test was applied when more than 10 specimens were attached (*: $0.01 < p < 0.05$; **: $0.001 < p < 0.01$; ***: $p < 0.001$; n.s.: not significant).

Table 2. Distribution of the attached larvae of *A. fontinalis* on dorsal and ventral sites of the abdomen of the most important chironomid hosts relative to host sex (dorsal and ventral attached larvae in %).

	Hosts	Mites total	Mites attached dorsally	Mites attached ventrally
Synorthocladius semivirens				
♂♂	65	138	44.9	42.0
♀♀	91	174	55.1	58.0
Rheocricotopus fuscipes				
♂♂	106	150	42.9	52.9
♀♀	89	137	57.1	47.1
Paratrichocladius skirwithensis				
♂♂	31	57	44.9	61.4
♀♀	29	49	55.1	38.6

nearly exclusively and *F. rouxi* preferentially attached to intersegmental sites of the host, whereas *A. nodipalpis* and *H. nigromaculatus* were both found on intersegmental sites and on the segments themselves. Another differentiation concerned the detailed attachment sites on the abdomen of the host: in nearly all mite species, the abdominal attachment sites were not equally occupied. Most larvae of *A. fontinalis* were attached between the third and fourth segments. Compared with *A. fontinalis*, most larvae of *A. nodipalpis* were attached more caudally, both intersegmentally and at the segment itself. Most larvae of *F. rouxi* were attached intersegmentally between segments 1/2 and 2/3. In *H. nigromaculatus*, nearly all specimens were attached to the first segment or between the first and second segments. The preferred sites of *H. nigromaculatus* were thus more caudal compared with those of *A. fontinalis* and *A. nodipalpis*. The order of attachment on the abdomen of the host, from anterior to posterior, was: *H. nigromaculatus*, *F. rouxi*, *A. fontinalis*, *A. nodipalpis*.

A comparison of the attachment sites of all species with each other (based on the summarized data) revealed that each species had different preferred sites (for all pairs of species $p < 0.001$).

The abdominal attachment of *A. fontinalis*, *A. nodipalpis*, *F. rouxi* and *H. nigromaculatus* on the most important chironomid host species is also illustrated in Table 1. At the species level, a similar preference for specific attachment sites was reflected as that previously seen for all specimens together. The sites of nearly all parasite–host associations were not equally distributed and there were preferences for selected sites.

For some mite species, specific attachment sites on the host abdomen varied between different host species. The possible pairs of hosts are compared with each other (Table 3). The attachment sites of *A. fontinalis* larvae on *S. semivirens* did not obviously differ from those on *R. fuscipes*, whereas there were differences in the preferred sites between *S. semivirens* and *P. skirwithensis* and between *R. fuscipes* and *P. skirwithensis*. Moreover, two pairs of host species of *A. nodipalpis* exhibited differences between attachment sites (*R. fuscipes* versus

Orthocladius sp. and *R. fuscipes* versus *P. convictum*). For other host species, no differences in sites could be demonstrated.

The previous analysis was based on the sum of the attachment sites of single parasitized (one mite per host) and intraspecific multiple parasitized hosts (more than one mite per host). Observation of hosts parasitized only by one larva, i.e. in the absence of infra- and interspecific competition for space, might have shown another distribution of the larvae; however, these larvae also had variations in attachment sites (Table 4). With regard to the exact attachment position, no distinct differences became evident when single-occurring larvae were compared with the pooled data for all larvae. The order of preferred attachment sites over the length of the abdomen of the host was similar to that for all mites. However, the larvae attached to most of the hosts had a non-homogeneous distribution. For some rare water mite species, the specificity of attachment sites was not statistically significant, probably because of the small numbers of specimens (e.g. *A. fontinalis* attached to *P. skirwithensis*). The preference for ventral or dorsal sites in the case of larvae of *A. fontinalis* attached to *R. fuscipes* and *S. semivirens*, as previously shown for all larvae (Table 1), was also reflected by single-attached larvae.

The statistical analysis between the abdominal attachment sites between hosts parasitized by one larva and multiple-parasitized hosts by the respective water mite species showed only slight differences. When the summarized data were compared (columns 'total' in Tables 1 and 4), only *A. fontinalis* larvae were significantly different ($0.001 < p < 0.01$). If one considers the differences in attachment sites between single- and multiple-parasitized hosts at the level of the host species (only parasite–host associations with more than 50 attached water mite larvae were considered), small differences between mites parasitizing hosts individually and those in total were possible, as was found for

Table 3. Comparison of the attachment sites of four water mite species on the most abundant hosts.

Aturus fontinalis	
S. semivirens vs. *R. fuscipes*	n.s.
S. semivirens vs. *P. skirwithensis*	***
R. fuscipes vs. *P. skirwithensis*	***
Atractides nodipalpis	
R. fuscipes vs. *Orthocladius* sp.	**
R. fuscipes vs. *P. convictum*	*
Orthocladius sp. vs. *P. convictum*	n.s.
Feltria rouxi	
P. skirwithensis vs. *Orthocladius* sp.	n.s.
Hygrobates nigromaculatus	
Microspectra spp. vs. *Tanytarsus* spp.	n.s.

For each water mite species, only pairs of abundant parasite–host associations were considered if the number of larvae was >50. The results for all segmental and intersegmental abdominal attachment sites were combined (see Material and methods). The asterisks indicate the significance of the χ^2-test: *: $0.01 < p < 0.05$; **: $0.001 < p < 0.01$; ***: $p < 0.001$; n.s.: not significant.

Table 4. Occurrence of the chironomid parasitizing water mites at the various abdominal segments of the body of the host.

		Aturus fontinalis				*Atractides nodipalpis*				*Feltria rouxi*				*H. nigromaculatus*		
		R. fuscipes	*S. semivirens*	*P. skirwithensis*	Total	*R. fuscipes*	*Orthocladius* sp.	*P. convictum*	Total	*Orthocladius* sp.	*P. skirwithensis*	*R. fuscipes*	Total	*Tanytarsus* spp.	*Micropsectra* spp.	Total
Abdomen	th/1										1		1			
Intersegmental	1/2	2	2	4	8		1		1	13	4	4	21	2	3	5
	2/3	33	11	6	50	3			3	10	4	3	17			
	3/4	88	59	6	153	2			2		2		2		1	1
	4/5	6	1	1	8	16			16							
	5/6	1		2	3	8	3	4	15	1			1			
	6/7					10	7	4	21							
	7/8					2		1	3							
		***	***	n.s.	***	***	n.s.		***	n.s.	n.s.		***			
Abdomen	1													23	16	39
Segmental	2									1		1	2			
	3					4			4							
	4					10	1		11							
	5					39	4		43							
	6					28	9		37							
	7					9	6	1	16	1			1			
	8					***	n.s.		***							
Larvae on abdomen		130	73	19	222	131	31	10	172	26	11	7	44	25	20	45
Ind. abd. dorsally (%)		1.5	71.2	36.8	27.5	3.8	6.5	90.0	9.3	26.9	9.1	14.3	20.5	0.0	5.0	2.2
Ind. abd. ventrally (%)		98.5	28.8	63.2	72.5	96.2	93.5	10.0	90.7	73.1	90.9	85.7	79.5	100.0	95.0	97.8
		***	***	n.s.	***	***	***		***	*	**		***	***	***	***

Only those parasite–host associations were considered in which one water mite larva was attached to one single host. The χ^2-test was applied when more than 10 specimens were attached (*: $0.01 < p < 0.05$; **: $0.001 < p < 0.01$; ***: $p < 0.001$; n.s.: not significant).

A. fontinalis larvae on *R. fuscipes* and *A. nodipalpis* larvae on *R. fuscipes* ($0.01 < p < 0.05$).

In a comparison of the attachment sites for host taxa parasitized by two water mite species (*R. fuscipes* and *Orthocladius* sp., considering only parasite–host pairs with more than 50 attached water mite larvae), the attachment of *A. fontinalis* larvae differed significantly from those of *A. nodipalpis* both for hosts carrying a single mite or for hosts parasitized by one or more mites ($p < 0.001$). *Atractides nodipalpis* and *F. rouxi* larvae also showed different attachment sites on *Orthocladius* sp. hosts (considering hosts parasitized by one or more mites, $p < 0.001$).

Discussion

Before the present investigation, only the studies of Efford (1963) and Ullrich (1978) had dealt with the species-specific attachment sites of stream-dwelling water mites on chironomid hosts and only Efford had examined the detailed position of the sites as presented here. Other than that, the tagmata of attachment sites had been reported for single stream living species (Smith and Oliver 1986; Smith et al. 2001).

In some parasite–host associations, the specificity for different attachment sites of the host is seen as being one of several strategies of host partitioning (Reilly and McCarthy 1993). This site-specific attachment might avoid intra- and interspecific competition. There is also evidence for host partitioning by different feeding sites in closely related species. For example, Lanciani (1970) has shown that a number of *Eylais* species use different attachment sites of their heteropteran hosts but little is known about intraspecific competition for different sites. In the following, competition for feeding sites on hosts will be revealed at different levels and other explanations than competition for the observed patterns of attachment will also be discussed.

Attachment to different host tagmata

The observed grouping of mites parasitizing the thorax (*P. eximia, Sperchon clupeifer, S. setiger, S. verrucosa, Lebertia inaequalis*) or abdomen (*A. fontinalis, A. nodipalpis, F. rouxi, H. nigromaculatus*) of the host agrees with data for the whole genera (Smith and Oliver 1986; Smith et al. 2001). Thus, this difference could be considered from a phylogenetic point of view: the attachment to thoracic sites can be viewed as the plesiotypic condition and the exploitation of abdominal sites as apotypical behaviour that brings advantages for the mites (Smith et al. 2001). Utilization of the abdomen facilitates the occurrence of more parasites on one host specimen. Additionally, abdominal parasites are often smaller than those attached to the thorax. Attachment to the abdomen of the host is apparently more stable than that to thorax, as exhibited by the numbers of detached thoracic parasites compared with abdominal parasites (Ullrich 1978; Smukalla and Meyer 1988, this study). Possibly, the presence or absence of stylostomes and/or differences in stylostome morphology contribute to the variability in attachment stability (see Smith 2003).

The preferred attachment to the thorax of the host as documented here has also been shown for *P. eximia* parasitizing Chironomidae by Ullrich (1978). Detailed sites have not been reported for the other thorax-attached species to date but *Sperchon setiger* also prefer the thorax of simuliids, the only observed host taxon in the studies of Ullrich (1978) and Gledhill et al. (1982). The attachment sites of *A. fontinalis* have hitherto been unreported, whereas those of the taxonomically vague *H. nigromaculatus* has been previously described by Martin and Davids (2002) in comparison with non-parasitic populations from standing waters. In the studies of Efford (1963) and Ullrich (1978), attachment to the abdomen of the host has also been found in *A. nodipalpis* and *F. rouxi*.

However, attachment site position may vary from host taxon to host taxon. For other host taxa, such a distinct difference may not even be present. For example, *P. eximia* larvae parasitizing Trichoptera attach to thoracic and to abdominal sites (Ullrich 1978; Martin 1998).

Specific attachment to thorax and abdomen of chironomid hosts

Wiles (1987) has shown that the larvae of a standing-water species of the genus *Hydrodroma* has a distinct specificity to a restricted area on the thorax of the host, probably because of the less marked sclerotization that enables the larvae to penetrate the cuticle in this region. The observed preference for the forelegs' coxae of *P. eximia* described here has also been noted by Ullrich (1978). As in the present study, Ullrich (1978) has found additional attachment to the region between head and thorax. The less pronounced preference for the forelegs by larvae of the other four thoracic parasites in the presented data is reflected by other Sperchontidae and *Lebertia* species in the studies of Efford (1963) and Ullrich (1978). The specificity for sites on the thorax of the host is probably in general based on such thinner regions of the integument.

The attachment of abdominal parasites shows an essential difference. The larvae of *A. fontinalis* and, with a few exceptions, of *F. rouxi*, are nearly exclusively attached intersegmentally to the soft integument between sclerotized segments. The simultaneous attachment of *A. nodipalpis* and *H. nigromaculatus* to segmental sites is probably caused by differences in locating the definite attachment site. On reaching an adequate chironomid pupa, larvae such as those of *A. fontinalis* and *F. rouxi* attach to the pupal skin without feeding, being only fixed by their legs. After waiting inactively, they become active once the imago starts to hatch, at which time they move from the skin of the pupa to that of the imago from which feeding starts. On the other hand, the larvae of *A. nodipalpis* and *H. nigromaculatus* contact the developing imago by embedding their palpal claws through the pupal integument. This has been described for the genera *Atractides* and *Hygrobates* (Ullrich 1978) and also for larvae of some *Unionicola* species (Böttger 1972b; Hevers 1978). During the course of the emergence of the chironomid imago from the pupal skin, these larvae are passively pulled through the pupal skin by the insect. Feeding does not start before this process is completed. This mode of obtaining a host may enable these larvae to attach to the subsequently more strongly sclerotized sclerites: the earlier attainment of the segment would enable the larvae to use their chelicerae before the segmental chitin has hardened. It is not clear from investigations of *Atractides*, *Hygrobates* or *Unionicola* whether this difference between segmental and intersegmental attachment sites has been neglected or is not apparent in the observed host species, e.g. in the studies of Efford (1963) or Kouwets and Davids (1984). In other studies, the observation of attachment sites is highly detailed and separates inter- and intrasegmental sites, but only at the genus level (e.g. Booth and Learner 1978; LeSage and Harrison 1980). However, the observations of Hevers (1978) for *Unionicola* species support the hypothesis that species that are able to attach to the host before the hatching of the imago regularly fix onto the sternites and tergites of abdominal segments.

The four abdominal parasites of the present study have different preferred attachment sites along the longitudinal direction of their hosts. *Hygrobates nigromaculatus* and *F. rouxi* attach more posteriorly than *A. fontinalis* and *A. nodipalpis*. This arrangement is similar for all individuals, independent of the host species or whether the mites occur alone or with other larvae (Tables 1 and 4). This kind of preference for distinct sites is also generally shown by other water mite taxa parasitizing chironomids (Efford 1963; Booth and Learner 1978; LeSage and Harrison 1980). A preference for extreme anterior attachment sites on the abdomen of the host by *H. nigromaculatus* agrees with observations made by Booth and Learner (1978) for the genus *Hygrobates* living in a body of standing water.

Interestingly, Efford (1963) partly dealt with the same species as investigated in the present study, viz. *A. nodipalpis* and *Feltria romijni* (subsequently synonymized with *F. rouxi* by Bader 1974) but he found no *A. nodipalpis* larvae attached to segmental sclerites. A comparison of attachment sites was made for these two species: in neither *A. nodipalpis* nor *F. rouxi* were identical attachment sites found (significant difference; χ^2-test; $p < 0.001$). This observation should be interpreted cautiously because, at least sometimes (see below), the specificity of the attachment site for the larvae of one water mite species was obviously related to the host species. A comparison at the level of the same host species was impossible, since, in the present study, the most abundant host species were different from those in the investigation of Efford (1963). Nevertheless, even in his study, *F. rouxi* displayed a slight preference for attachment sites located anterior to those preferred by *A. nodipalpis*.

In spite of the general pattern of attached mites, there is evidence that the exact attachment sites of one mite species are different between host species (*A. fontinalis*, *A. nodipalpis*; Table 3). This is astonishing, since the host species (with the exception of *P. convictum*) belong to the same chironomid subfamily

Orthocladiinae. Although the morphological differences between these species might not be pronounced, small differences in morphology, behaviour or other features seem to lead to slightly differences in attachment sites. Moreover, Booth and Learner (1978) have pointed out that differences in host morphology of chironomids could cause differences in the patterns of attachment.

In summary, the specific attachment sites within the tagmata of the host show a species-specific pattern that seems to be relatively independent of the host species.

Is there competition for attachment sites?

The premise that mites are able to avoid interspecific competition by utilizing different sites for parasitism is reflected by the results given here. However, this avoidance manifests itself only if an individual host is parasitized by different mite species. In the present study of 978 parasitized chironomids, only 63 were parasitized by more than one water mite species (=6.4%, Martin, in prep.). Thus, the reduction of competition that is theoretically possible does not play an important role in the implementation of parasite–host associations of the observed stream. Nevertheless, in all 17 chironomids that are simultaneously parasitized by *P. eximia* and larvae of other species, *P. eximia* larvae attach to the thorax, whereas the other species attach to the abdomen. In other studies, multiple parasitism by different mite species seems to be more common. Efford (1963) found many chironomids that were simultaneously parasitized by *F. rouxi* and *A. nodipalpis* and even showed changes in preferred sites for these species: if *F. rouxi* is present, *A. nodipalpis* larvae take up a more posterior position on the abdomen of the most important *Stempellinella* host. At first sight, the specificity of attachment site seems to be a flexible pattern, but we have also to consider the different modes of host attainment (see above). Efford supposed that *Feltria* larvae have advantages in attaching to more anterior positions on the pupal case of the host than have those of *Atractides* because the former are smaller and the space between the pupa and its case becomes more narrow anteriorly. However, this would mean that the attachment of *Atractides* larvae by their chelicerae before the imago emerges from the pupal case (see Ullrich 1978) occurs relatively shortly before ecdysis. Alternatively, *Atractides* larvae would have advantages over *Feltria* larvae, which do not attach their mouthparts before the imago emerges. Nevertheless, the detailed course of competition between larvae, inter- and also intraspecifically, is difficult to observe.

Evidence has been provided for intraspecific competition for space not only from species in standing water (Lanciani 1976), but also in stream living species: Davies (1959) demonstrated that the larvae of *Sperchon jasperensis* Marshall 1929 were smaller when many had infested a single host than when few were present. In the present study, the analysis of the parasite–host association in which only one mite individual is involved has provided surprising data. In single mites, the bias towards one exactly preferred attachment site was expected to be stronger. Indeed, they also showed a relatively wide variation in occupied sites but, in both cases, the larvae were most abundant at the same sites. In consequence, a preference for a distinct segment seems to be improbable. The sites where most larvae of a species are attached may reflect those sites that are first reached by the mite larvae after the emergence of the host imago. Thus, the pattern of larval distribution on the host shows the probability of the successful attainment of a certain site and is therefore variable in a restricted frame. If more than one larva tries to attach to one host, they may find it easier to attach to neighbouring sites than to struggle with intraspecific or an interspecific competitors.

In conclusion, evidence has been provided that stream-dwelling water mite larvae have species-specific patterns of distribution on the body of the host. Interspecific competition is probably more severe than intraspecific competition for niche partitioning. A type of intrinsic site specificity seems to be involved that is probably based on host morphology and early parasite behaviour. Gledhill et al. (1982) has noted that pupal morphology of simuliids determines the attachment of the larvae of *Sperchon setiger*. This is also probably true for chironomid pupae and, together with the detailed attainment strategy towards the host imago, is a more important feature in successful attachment sites than direct competition between single larvae.

Additional influences on the observed attachment patterns

As far as is known, all water mite females lay eggs subaquatically. For larvae, the position of the egg or the clutch can be an important factor for later life. In principle, three ways of finding a host are undertaken by water mites: (1) the host and the parasite stay in the water (e.g. mites parasitizing insect larvae; not seen in the species of the present study), (2) mite larvae have to leave the water to find a host (*P. eximia* in the present study) or (3) water mites use insect pupae phoretically and feed on the host imago only after ecdysis (seen in all other species in the present study). In any case, most water mite larvae must locate a host within a few days as a result of their short life expectancy (Smith 1988). Moreover, for nearly all the species observed here, laboratory-reared larvae have a maximum life duration without parasitism of 2–5 weeks (Martin 1998). If the larvae find a host in that period, the successful attachment for feeding depends on the host biology. With the exception of *P. eximia*, which is a representative of a 'terrestrial mite' (Mitchell 1957) and searches for a host above water, all species in the present study have to find a suitable chironomid pupa. Hosts are not defenceless against attacking mites, e.g. nematocerous pupae show vigorous tumbling (mosquitoes; Smith and McIver 1984, chironomids; Ullrich 1978; Wiles 1987, own observations). Thus, the final attachment or feeding sites reflect not only a real preference for a site, but also grooming by the host, competitive exclusion or some combination of events.

Competition for attachment sites is different for terrestrial and aquatic mite larvae. *Protzia eximia* contacts its host by crawling and/or even by jumping from the land or from the water surface (Jones 1967; Ullrich 1978). Terrestrial mites often have no distinct specificity of attachment sites. Although *P. eximia* prefers the prothorax as an attachment site on chironomid hosts, the attachment on Trichoptera varies from thorax to abdomen (Ullrich 1978; Martin 1998). Terrestrial mite larvae have relatively more time to search and find a suitable place than do aquatic larvae. They can afford to have more than one attempt to find a suitable host. Intraspecific competition may play a role in cases of high densities of the mite species in the respective habitat. Then, nearly all possible sites are used, e.g. the larvae of *Limnochares aquatica* (Linnaeus, 1758), which sometimes achieve very high intensities on their heteropteran hosts (Böttger 1972a). *Protzia eximia* is a relatively rare species in the Farver Au and, since it is the only terrestrial parasite in the stream that parasitizes chironomids, neither intra- nor interspecific competition will be very important. If the larvae of *P. eximia* meet hosts that are already parasitized by other aquatic larvae, then they have to use other sites for feeding.

Aquatic larvae, on the other hand, have different problems in reaching a host and possibly try avoid competition. They are under higher pressure to attach successfully to a suitable host, since they probably have few chances of repeatedly finding a host. In addition, many of these species produce few eggs and thus the female probably has to take great care in choosing a place for ovipositioning in order to enable the hatching larvae to reach a host. In the laboratory, I have observed that females of *A. fontinalis* and *F. rouxi* seldom lay eggs but oviposit more easily when one egg is already present. The presence of an egg may influence egg-laying behaviour of conspecific females (Martin 1998). Females may prefer to lay eggs in the vicinity of potential hosts and thus ensure a higher probability of host finding. Most aquatic water mite larvae that have found a suitable chironomid pupa do not start to feed immediately but merely grasp the pupal skin by the legs and wait inactively (Ullrich 1978). Once the chironomid imago emerges from the pupal skin, the mites becomes active and moves to the imago at the moment of ecdysis. As a result, intra- and interspecific competition can occur at two specific times: during attachment to the pupal skin and once the definitive attachment and feeding site is reached. Mitchell (1998) has demonstrated that, for ceratopogonid hosts, the probability of arousal for the larvae increases with the number of mites on a pupa. In contrast, the chances of larvae transferring from the pupa to a midge are reduced by interference among larvae in large groups.

In some cases, differences in dorsal/ventral attachment of the larvae can be explained by the sex of the host. For example, Mitchell (1959) concluded that, for odonates parasitized by *Arrenurus* species, the site that each species selected on the host was directly related to the egg-laying strategy of the female odonate. However, in later studies of odonate parasites, he observed that the larvae attached to the first membrane that they encountered as they moved from the pupal exuvium (Mitchell 1968). The four species in the present study preferentially parasitizing the abdomen of the host favour ventral attachment sites if the overall distribution is considered without discrimination of the host species. If preferences at the level of host species are taken into account, for most species, a preference for ventral attachment or no preference can be shown. This is surprising, since in the study of Efford (1963), only a few larvae were attached ventrally. A ventral attachment could be seen as an adaptation to a better detachment from the host. It is possible that the larvae can more easily recognize their vicinity to water when attached ventrally. Ventral attachment sites could also be seen as advantageous for larvae parasitizing host females, since these often lay their eggs sub-aquatically and, thus, the detachment of larvae is facilitated. However, there is no strong sex-biased relationship towards preference for female hosts in the present study (own unpublished data). In contrast to most species of the present study, the larvae of *A. fontinalis* preferably attach to dorsal sites on individuals of *S. semivirens*. For the larvae of *A. fontinalis*, the patterns of dorsal/ventral attachment is independent of the sex of the host (Table 2); thus, the egg-laying behaviour of the female host obviously plays no role in the attachment of ventral or dorsal attachment sites. Moreover, dorsal/ventral attachment of the larval water mites seems to be connected with host morphology.

Outlook

Specificity for attachment sites is seen as one possibility of reducing competition in water mites and thus of enabling the coexistence of water mite larvae of different species on one host species (Lanciani 1970; Reilly and McCarthy 1993). Other potential strategies that might avoid concurrence are the parasitization of different host species, the different seasonality of the larvae or the different exploitation of the sex of the host. Certainly, being adapted to a specific site should increase the fitness of the parasite in a particular site over the fitness of the parasite at some other site (Bush et al. 2001). However, attachment site selectivity can not be seen as the only possibility for host partitioning. For the parasite–host associations of the species observed here, parasitological data and differences in the life history strategies of the water mites will be presented elsewhere (Martin, in prep.). Moreover, in these species, strategies for host partitioning are utilized other than attachment site differentiation. An example is the host utilization of *R. fuscipes* by *A. fontinalis* and *A. nodipalpis* in the Farver Au. The first peak of emergence of this chironomid in spring is nearly exclusively utilized by *A. nodipalpis*, whereas the later emerging *R. fuscipes* specimens are parasitized by *A. fontinalis* (own unpublished data).

For a water mite species, the preferred attachment sites might exhibit slight differences between different host species. Therefore, the attachment sites of a water mite species at different localities with different host inventories will also differ. Further comparative studies should be of interest dealing with more numerous study sites in different catchment areas harbouring similar sets of species of parasites and hosts.

Acknowledgements:

I thank Dr Claus-Joachim Otto (Fahrenkrug, Germany) for the determination of the chironomids. Dr Amke Caliebe (University of Kiel) helped with the statistics.

References

Bader C. 1974. Zur Revision holländischer Hydrachnellae (Acari). I. Feltriidae. Entomol. Berichten 34: 149–152.

Böttger K. 1972a. Vergleichend biologisch-ökologische studien zum Entwicklungszyklus der Süßwassermilben (Hydrachnellae, Acari). I. Der Entwicklungszyklus von *Hydrachna globosa* und *Limnochares aquatica*. Int. Rev. Ges. Hydrobiol. 57: 109–152.

Böttger K. 1972b. Vergleichend biologisch-ökologische Studien zum Entwicklungszyklus der Süßwassermilben (Hydrachnellae, Acari). II. Der Entwicklungszyklus von *Limnesia maculata* und *Unionicola crassipes*. Int. Rev. Ges. Hydrobiol. 57: 263–319.

Booth J.P. and Learner M.A. 1978. The parasitization of chironomid midges (Diptera) by water-mite larvae (Hydrachnidia: Acari) in a eutrophic reservoir in South Wales. Arch. Hydrobiol. 84: 1–28.

Bush A.O., Fernandez J.C., Esch G.H. and Seed J.R. 2001. Parasitism, The Diversity and Ecology of Animal Parasites. Cambridge University Press, pp. 566.

Davids C., Nielsen G.J. and Gehring P. 1977. Site selection and growth of the larvae of *Eylais discreta* KOENIKE, 1897 (Acari, Hydrachnellae). Bijdr. Dierk. 46: 180–184.

Davies D.M. 1959. The parasitism of black flies (Diptera, Simuliidae) by larval water mites mainly of the genus *Sperchon*. Can. J. Zool. 37: 353–369.

Efford I.E. 1963. The parasitic ecology of some watermites. J. Anim. Ecol. 32: 141–156.

Gledhill T., Cowley J. and Gunn R.J.M. 1982. Some aspects of the host : parasite relationships between adult blackflies (Diptera; Simuliidae) and larvae of the water mite *Sperchon setiger* (Acari; Hydrachnellae) in a small chalk stream in southern England. Freshwater Biol. 12: 345–357.

Hevers J. 1978. Interspezifische Beziehungen zwischen *Unionicola*-Larven (Hydrachnellae, Acari) und Chironomidae (Diptera, Insecta). Ver. Gesell. Ökologie 7: 211–217.

Jones R.K.H. 1967. Descriptions of the larvae of *Aturus scaber* Kramer, *Protzia eximia* Protz and *Piona uncata* Koenike with notes on the life-histories of the later two. [Hydracarina]. Ann. Limnol. 3: 231–247.

Kouwets F.A.C. and Davids C. 1984. The occurrence of chironomid imagines in an area near Utrecht (The Netherlands), and their relation to water mite larvae. Arch. Hydrobiol. 99: 296–317.

Lanciani C.A. 1970. Resource partitioning in species of the water mite genus *Eylais*. Ecology 51: 338–342.

Lanciani C.A. 1976. Intraspecific competition in the parasitic water mite, *Hydryphantes tenuabilis*. Am. Midland Nat. 96: 210–214.

LeSage L. and Harrison A.D. 1980. The biology of *Cricotopus* (Chironomidae: Orthocladiinae) in an algal-enriched stream: part II. Effects of parasitism. Arch. Hydrobiol. /Suppl. 58: 1–25.

Margolis L., Esch G.W., Holmes J.C., Kuris A.M. and Schad G.A. 1982. The use of ecological terms in parasitology (report of an ad hoc committee of the American Society of Parasitologists). Jo. Parasitol. 68: 131–133.

Martin P. 1996. Faunistisch-ökologische Benthosstudien an den Wassermilben (Hydrachnidia, Acari) zweier Bäche des Norddeutschen Tieflandes (Ostholsteinisches Hügelland, Schleswig-Holstein). Faunistisch-Ökologische Mitteilungen 7: 153–167.

Martin P. 1998. Zur Autökologie der Wassermilben (Hydrachnidia, Acari) zweier norddeutscher Tieflandbäche. University of Kiel, PhD thesis, pp. 269.

Martin P. 2000. Larval morphology and host-parasite associations of some stream living water mites (Hydrachnidia, Acari). Arch. Hydrobiol. /Suppl. 121: 269–320.

Martin P. and Davids C. 2002. Life history strategies of *Hygrobates nigromaculatus*, a widespread palaearctic water mite (Acari, Hydrachnidia, Hygrobatidae). In: Bernini F., Nannelli G., Nuzzaci G. and de Lillo E. (eds), Acarid Phylogeny and Evolution. Adaptations in mites and ticks. Kluwer Academic Publishers, Dordrecht, The Netherlands, pp. 101–110.

Meyer E. 1985. Der Entwicklungszyklus von *Hydrodroma despiciens* (O. F. Müller 1776) (Acari: Hydrodromidae). Arch. Hydrobiol. /Suppl. 66, 13: 321–453.

Mitchell R. 1957. Major evolutionary lines in water mites. Syst. Zool. 6: 137–148.

Mitchell R. 1959. Life histories and larval behavior of Arrenurid water-mites parasitizing Odonata. J. New York Entomol. Soc. 67: 1–12.

Mitchell R. 1968. Site selection by larval water mites parasitic on the damselfly *Cercion hieroglyphicum* Brauer. Ecology 49: 40–47.

Mitchell R. 1998. The behavior of *Arrenurus* larvae (Acari: Hydrachnidea) parasitizing Diptera. Acarologia 39: 49–55.

Proctor H. and Pritchard G. 1989. Neglected predators: water mites (Acari: Parasitengona: Hydrachnellae) in freshwater communities. J. N. Am. Benthol. Soc. 8: 100–111.

Reilly P. and McCarthy T.K. 1993. Attachment site selection of *Hydrachna* and *Eylais* (Acari: Hydrachnellae) watermite larvae infecting Corixidae. J. Nat. Hist. 27: 599–607.

Sachs L. 1993. Statistische Methoden – Planung und Auswertung. Springer-Verlag, Berlin, pp. 312.

Smith B.P. 1988. Host-parasite interaction and impact of larval water mites on insects. Ann. Rev. Entomol. 33: 487–507.

Smith B.P. 2003. Diversity of stylostome structure among parasitic larval water mites (Prostigmata: Hydrachnida). In: Smith I.M. (ed.), An Acarological Tribute to David R. Cook – From Yankee Springs to Wheeny Creek. Indira Publishing House, West Bloomfield, Michigan, pp. 239–255.

Smith B.P. and McIver S.B. 1984. The pattern of mosquito emergence (Diptera: Culicidae; *Aedes* spp.): their influence on host selection by parasitic mites (Acari: Arrenuridae; *Arrenurus* spp). Can. J. Zool. 62: 1106–1113.

Smith I.M. and Oliver D.R. 1986. Review of parasitic associations of larval water mites (Acari: Parasitengona: Hydrachnidia) with insect hosts. Can. Entomol. 118: 407–472.

Smith I.M., Cook D.R. and Smith B.P. 2001. Water mites (Hydrachnida) and other arachnids. In: Thorp J.H. and Covich A.P. (eds.), Ecology and Classification of North American Freshwater Invertebrates, Chapter 16: 2nd edition. Academic Press, San Diego, pp. 551–659.

Smukalla R. and Meyer E. 1988. Insect emergence from a shallow southern West German lake, with special reference to the parasitic host-associated water mite larvae. Hydrobiologia 169: 149–166.

Ullrich F. 1978. Biologisch-ökologische Studien an den Larven rheophiler Wassermilben (Hydrachnellae, Acari), Schlitzer Produktionsbiologische Studien (29). Arch. Hydrobiol. /Suppl. 54: 189–255.

Wiles P.R. 1987. Observations on the parasitic biology of the water mite *Hydrodroma despiciens pilosa* BESSELING (Acari: Hydrodromidae). Arch. Hydrobiol. /Suppl. 76: 369–392.

Experimental and Applied Acarology **34:** 113–125, 2004.

Communication via sex pheromones within and among *Arrenurus* spp. mites (Acari: Hydrachnida; Arrenuridae)

BRUCE P. SMITH* and JOY FLORENTINO
*Biology Department, Ithaca College, 953 Danby Road, Ithaca, NY 14850-7278, USA; *Author for correspondence (e-mail: smithb@ithaca.edu; phone: +1-607-274-3971; fax: +1-607-274-1131)*

Key words: Arrenuridae, Mating behaviour, Sex pheromone, Water mite

Abstract. We present direct experimental evidence of pheromone use in six species of *Arrenurus* and indirect evidence for four species, including members of the subgenera *Megaluracarus*, *Truncaturus*, and *Arrenurus*. Water in which females were housed elicited arrestant behaviour in males, males oriented to the source, and at least some individuals in each species assumed the male readiness posture, a precursor to coupling. Most species responded to water treated with conspecific females, but there was also interspecific sex pheromone responsiveness. *Arrenurus manubriator* and *A. megalurus* demonstrated reciprocal pheromone cross-attractancy. Males of *A. major*, *A. marshallae*, and *A. birgei* responded to water from females of related species from within their subgenera. *Arrenurus apetiolatus* males failed to respond to conspecific female-treated water, but the same water elicited arrestant behaviour and orientation in *A. manubriator*. Heterospecific reactions to female-conditioned water were limited to cases involving members of the same species group and were not seen between species representing different species groups or different subgenera. The species for which cross-attractancy has been demonstrated commonly co-occur in nature, so apparently these pheromones are of limited value for species recognition. Shared reaction to sex pheromones provides additional evidence for inferring close phylogenetic relationship among species, and thus far, corresponds with morphological evidence based on adult males and larvae.

Introduction

Recently, Smith and Hagman (2002) demonstrated that in at least one species of *Arrenurus* water mite, females produce a male-attractant stimulus that can be extracted from water – presumably a pheromone. When exposed to this stimulus, males stop swimming (arrestant behaviour) and either remain motionless or crawl slowly towards the source, and if strongly stimulated, assume a characteristic readiness posture (Proctor and Smith 1994) with the fourth legs held against the dorsum and bent at the genuotibial joint. Males of various other *Arrenurus* spp. have been reported to exhibit this posture (e.g., Böttger 1962; Proctor and Wilkinson 2001). Given that various species share the male readiness position, it is quite possible that those species also use sex pheromones.

Sonenshine (1985) reviewed chemical communication in Acari, and noted that many species use the same or similar chemicals, which is in strong contrast to the species-specificity seen in many insects, especially Lepidoptera (e.g., Löfstedt 1993). Much of the specificity in insects is based upon species responding only to relatively precise blends of isomers, sometimes requiring small quantities of secondary components (e.g., Silk and Kuenen 1988; Löfstedt 1993; Roelofs 1995; Mori and Kuwahara 2000). Our impression from Sonenshine (1985) and subsequent papers is that sex pheromones for many species of Acari are single compounds or simple mixtures.

The genus *Arrenurus* is species-rich: there are over 900 described species (Viets 1987) and it is not difficult to discover undescribed species in most parts of the world. Given the species diversity and the relative simplicity of many acarine sex pheromones, it is quite probable that there is substantial interspecific sex pheromone responsiveness. We have two aims in our study: (1) to investigate whether female-emitted male-attractant sex pheromones are widespread among *Arrenurus* species, and (2) to explore species-specificity of the pheromones used.

Materials and methods

Collection and maintenance of mites

Most of the mites used in this study were collected from three field sites in Southern Ontario, Canada (Table 1). The Black River and Salmon River were sampled at their junctions with Highway 7 (ca. 10 km east of Madoc, and ca. 20 km east of Kaladar, respectively). Telephone Bay of Lake Opinicon, located approximately 6 km south-west of Chaffeys Lock, ca. 50 km north-east of Kingston, was also a source of experimental animals. Collections were made following the methods of Smith et al. (2001) using D-frame nets in waters of 0.25–1.5 m depth. Adult *Arrenurus* (*Megaluracarus*) *manubriator* Marshall were available from a sixth generation laboratory colony established from inseminated females previously collected from Telephone Bay of Lake Opinicon. A fifth generation colony of *Arrenurus* (*Truncaturus*) *rufopyriformis* Habeeb was another source of experimental animals, having originated from females collected from Mèr Bleue Bog (located on the south-eastern edge of Ottawa, Ontario, by Blackburn Hamlet). Both of these species are atypical of parasitengonine mites in that they forego larval feeding and any association with insect hosts (Smith 1998) which makes it feasible to maintain reproducing colonies in the laboratory. Adults of an undescribed species (*Arrenurus* (*Arrenurus*) new species, near *reflexus*) were raised from engorged larvae removed from parasitized adult *Leucorrhinia frigida* Hagen dragonflies netted at Hebert Bog (located ca. 14 km south-west of Chaffeys Lock on the Opinicon Road) in early July, 1995.

Field-collected mites were stored unfed and in darkness at 15 °C until about 1 week before being used in experiments. They were then warmed to and kept at room temperature (range 21–26 °C) under ambient fluorescent lighting

Table 1. Localities and dates for field-collected *Arrenurus* spp. mites used in this study. Subgenera '(*A.*)': *Arrenurus*, ' (*M.*)': *Megaluracarus*.

Lake Opinicon, September 2, 1995
A. (*A.*) *americanus* Marshall
A. (*A.*) *falcicornis* Marshall
A. (*A.*) *fissicorniformis* Cook
A. (*A.*) *magnicaudatus* Marshall
A. (*M.*) *apetiolatus* Piersig
A. (*M.*) *birgei* Marshall
A. (*M.*) *intermedius* Marshall
A. (*M.*) *megalurus* Marshall
Salmon River, September 3, 1995
A. (*A.*) *americanus* Marshall
A. (*A.*) *fissicornis* Marshall
A. (*A.*) *flabellifer* Marshall
A. (*A.*) *gennadus* Cook
A. (*A.*) *major* Marshall
A. (*A.*) *pseudosuperior* Cook
A. (*A.*) *superior* Marshall
A. (*M.*) *apetiolatus* Piersig
A. (*M.*) *birgei* Marshall
A. (*M.*) *megalurus* Marshall
Black River, September 3, 1995
A. (*A.*) *flabellifer* Marshall
A. (*A.*) *magnicaudatus* Marshall
A. (*M.*) *marshallae* Piersig
A. (*M.*) *megalurus* Marshall
A. (*M.*) new species, near *unisinuatus*

(ca. 6–10 h lighting per day) and fed. Mites of most species were provided an excess of living ostracods from laboratory colonies. Adult *Arrenurus (Arrenurus) pseudosuperior* Cook and *Arrenurus (Arrenurus) superior* Marshall were maintained on small immatures of *Daphnia magna* Straus. Colonies of *A. manubriator* and *A. rufopyriformis* were kept in polypropylene containers, continually in room conditions, with an abundance of ostracods. Tap water aged for at least 24 h was used for housing mites and for experiments.

Experimental methods

Female-conditioned water was produced by storing female mites of a given species in a glass beaker (1 ml of water per mite) for 24 h at room temperature. Food was withheld during this period to avoid potential kairomones from prey. Water was conditioned using as many females as were available up to a maximum of 250. A comparable volume of water destined for use as a control was also stored in a glass beaker at room temperature for 24 h. Control and female-conditioned water was used in experiments immediately after the 24 h period. In both cases, beakers were kept covered with parafilm.

The design was to simultaneously test a group of males of a given species, using two sources of water: female-conditioned and an untreated control. The test arena was a glass petri dish (9 cm in diameter, 1.5 cm deep) containing 30 ml of 24-h aged tap water. The dish had a line drawn down the middle of the underside, thus dividing it into two regions. Two syringes were suspended opposite each other equidistant from the midline of the test arena and each about 2.5 cm from the dish's edge, with the needle tips just below the water's surface. One syringe contained control water, the other had female-conditioned water. During a test, the plungers of the two syringes were removed simultaneously so that the water in the two syringe barrels would drain by gravity. Responses by males were scored immediately after the syringes had finished draining. Only males exhibiting arrestant behaviour were counted, and we specifically tested whether male arrestant behaviour was oriented towards the putative pheromone (i.e., number of males on the side with the female-treated water vs. the number on the control side of the arena). We compared counts of arrested males from the two sides using a one-tailed sign test of association: the null hypothesis was that there would be no apparent orientation of stationary mites (non-significant, or negative result), the alternative hypothesis was that arrestant males would cluster on the side corresponding to the female-conditioned water (a positive result, being statistically significant). This criterion was used because males could stop swimming for other reasons besides an arrestant response to a pheromone source. Counting males that assumed the readiness posture would be a more definitive test, but it was impossible to accurately score groups of males tested simultaneously because individuals frequently hold the posture for only a fraction of a second to a couple of seconds, and it was too difficult to reliably recognize the posture without magnification. We found that males tested as groups were no more likely to show arrestant behaviour than males tested individually, and testing in groups allowed for larger sample sizes with less time and effort. However, we also noted that arrested males are more likely to exhibit the readiness posture after a collision with another mite, as also reported by Böttger (1962).

Given that the only way to demonstrate the presence of pheromone was to see a response by males, it was necessary to elicit an arrestant behaviour in males oriented to the source of female-conditioned water as a postive control. Pheromone production drops off dramatically when females are not well-fed or are in poor condition, and similarly, males will only exhibit an arrestant response when they are well-fed and in good condition. A positive result from testing males with conspecific female-conditioned water was an optimal control for the males' responsiveness to stimuli and for the females' production of pheromone. A test in which males responded to water from heterospecific females was still informative even with negative results from the controls, but negative results from heterospecific tests without evidence for male receptiveness and the presence of pheromone were ambiguous. Species without some positive evidence for pheromone use were dropped from the study.

A series of tests were run in each comparison, with female-conditioned water first being presented to conspecific males to test whether this species used female-emitted sex pheromones (and whether the males and females were in adequate condition). After a test, males were placed into a new petri dish with fresh water and allowed to sit for 15 min before the next test was conducted. After males had first been tested with conspecific female-conditioned water, the same males were tested with a sequence of water conditioned by females of other *Arrenurus* species. As many males as available up to a maximum of 40 per species were used in each comparison. Seven sets of comparisons were conducted in total, given that it was not possible to conduct tests with a wide range of species combinations within a reasonable period of time. All tests within a set were conducted within a two day period, including repeats of conspecific tests. Consequently, results of sets of comparisons are presented separately, and certain reference species are included in several different sets. The first set was repeated as confirmation, given that it was the first evidence for heterospecific responsiveness to pheromones among *Arrenurus* species; we decided to conduct other sets of tests to further explore the scope of cross-species reactivity to pheromones after having made this discovery. In all but one series of tests, 3 ml syringes were used each with 3 ml of solution, and it took approximately 75 s for the syringes to drain; in the fifth set of comparisons, 1 ml syringes and 1 ml of each solution were used and draining took about 25 s. Syringes used for female-treated water from a given species of mite were used for that species throughout a set of comparisons, and new syringes were used for each set of comparisons.

Results

There was no evidence of pheromone communication in nine species and they were omitted from further consideration. Male *Arrenurus* (*Megaluracarus*) *wardi* Marshall and *Arrenurus* (*Megaluracarus*) new species, near *unisinuatus* were tested with water from their own females and with water from the other species, but none showed a response. Females could not be reliably identified or were too few for *Arrenurus* (*Megaluracarus*) *intermedius* Marshall, *Arrenurus* (*Arrenurus*) *fissicornis* Marshall, *Arrenurus* (*Arrenurus*) *falcicornis* Marshall, *Arrenurus* (*Arrenurus*) *magnicaudatus* Marshall, *Arrenurus* (*Arrenurus*) *flabellifer* Marshall, *Arrenurus* (*Arrenurus*) *fissicorniformis* Cook, and *Arrenurus* (*Arrenurus*) *gennadus* Cook. Males of these species failed to respond to heterospecific female-conditioned water and were dropped from the study. We only had one male of *A. superior* and no females, but we nevertheless tested it with water conditioned by female *A. pseudosuperior*: the male arrested and assumed the readiness posture.

In the first set of comparisons (Table 2) we concentrated on species of the subgenus *Megaluracarus*, namely various combinations of *A. manubriator*, *Arrenurus megalurus* Marshall, *Arrenurus birgei* Marshall, and *Arrenurus marshallae* Piersig. Females of *A. marshallae* could not be reliably identified and only males were used in the study. This comparison was repeated, forming two trials. Male *A. manubriator* and *A. megalurus* responded and oriented to water from their own and each other's females in both trials (Table 2). Almost all arrested males would cluster on the side of the arena near the test syringe, and more distant males would crawl slowly towards the source. Water conditioned with female *A. manubriator* did not elicit a response from *A. marshallae* males, while water from *A. megalurus* females did cause *A. marshallae* males to arrest and orient (statistical significance in trial 2, $p = 0.059$ in trial 1; Table 2). Male *A. birgei* did not respond to conspecific stimuli, but responded to *A. manubriator* and *A. megalurus* female water: few oriented in trial 1 and neither case was statistically significant, but there were significant responses by *A. birgei* males to water from females of both species in trial 2 (Table 2).

Table 2. Tests for pheromone response as indicated by male arrestant behaviour among species of the subgenus *Megaluracarus*.

No. of males	Species of males	Female water	No. of males at test	No. of males at control	Sign test
Trial 1					
40	*A. manubriator*	*A. manubriator*	23	0	$p < 0.001$
40	*A. megalurus*	*A. megalurus*	26	2	$p < 0.001$
40	*A. megalurus*	*A. manubriator*	18	3	$p < 0.001$
40	*A. manubriator*	*A. megalurus*	13	2	$p < 0.005$
40	*A. marshallae*	*A. manubriator*	3	1	$p = 0.313$
40	*A. marshallae*	*A. megalurus*	11	4	$p = 0.059$
40	*A. birgei*	*A. birgei*	1	0	$p = 0.500$
40	*A. birgei*	*A. manubriator*	4	0	$p = 0.063$
40	*A. birgei*	*A. megalurus*	1	0	$p = 0.500$
Trial 2					
40	*A. manubriator*	*A. manubriator*	32	0	$p < 0.001$
16	*A. megalurus*	*A. megalurus*	12	1	$p < 0.005$
16	*A. megalurus*	*A. manubriator*	9	0	$p < 0.005$
40	*A. manubriator*	*A. megalurus*	29	0	$p < 0.001$
38	*A. marshallae*	*A. manubriator*	1	0	$p = 0.500$
38	*A. marshallae*	*A. megalurus*	7	0	$p < 0.008$
40	*A. birgei*	*A. birgei*	0	0	$p = 1.000$
40	*A. birgei*	*A. manubriator*	10	0	$p < 0.001$
40	*A. birgei*	*A. megalurus*	5	0	$p < 0.032$

A group of males of a given species were given simultaneous presentation of female-conditioned water of either the same or a different species and of control water. The number showing arrestant behaviour in the half of the test arena by the source of female-treated water and control water were recorded. Data were tested using a one-tailed sign test. The two trials were conducted in different weeks and hence are kept separate.

The second set of comparisons centered on testing whether *Arrenurus (Megaluracarus) apetiolatus* Piersig would respond to *A. manubriator* and *A. megalurus*. Males of *A. manubriator* and *A. megalurus* responded to water conditioned with conspecific females, but *A. apetiolatus* males did not (Table 3). We continued testing *A. apetiolatus* males with heterospecific-conditioned water and there was a significant arrestant and orientation response in two trials with water from *A. manubriator* females but no response with that from *A. megalurus* (Table 3).

Table 3. Tests for pheromone response as indicated by male arrestant behaviour among species of the subgenus *Megaluracarus* (see Table 2).

No. of males	Species of males	Female water	No. of males at test	No. of males at control	Sign test
40	*A. manubriator*	*A. manubriator*	27	0	$p < 0.001$
40	*A. megalurus*	*A. megalurus*	31	0	$p < 0.001$
38	*A. apetiolatus*	*A. apetiolatus*	1	1	$p = 0.500$
40	*A. manubriator*	*A. apetiolatus*	10	1	$p < 0.006$
40	*A. manubriator*	*A. apetiolatus*	15	0	$p < 0.001$
40	*A. megalurus*	*A. apetiolatus*	2	1	$p = 0.500$

Table 4. Tests for pheromone response as indicated by male arrestant behaviour among species of the subgenus *Arrenurus* (see Table 2).

No. of males	Species of males	Female water	No. of males at test	No. of males at control	Sign test
40	*A. americanus*	*A. americanus*	27	2	$p < 0.001$
19	*A.* n.sp. nr. *reflexus*	*A.* n.sp. nr. *reflexus*	10	1	$p < 0.006$
14	*A. pseudosuperior*	*A. pseudosuperior*	13	1	$p < 0.001$
40	*A. americanus*	*A.* n.sp. nr. *reflexus*	0	0	$p = 1.000$
14	*A. pseudosuperior*	*A.* n.sp. nr. *reflexus*	0	0	$p = 1.000$
19	*A.* n.sp. nr. *reflexus*	*A. pseudosuperior*	0	0	$p = 1.000$
40	*A. americanus*	*A. pseudosuperior*	0	0	$p = 1.000$
14	*A. pseudosuperior*	*A. americanus*	3	2	$p = 0.500$
19	*A.* n.sp. nr. *reflexus*	*A. americanus*	1	0	$p = 0.500$

Table 5. Tests for pheromone response as indicated by male arrestant behaviour among species of the subgenus *Arrenurus* (see Table 2).

No. of males	Species of males	Female water	No. of males at test	No. of males at control	Sign test
40	*A. americanus*	*A. americanus*	25	2	$p < 0.001$
24	*A. major*	*A. americanus*	18	0	$p < 0.001$
12	*A. pseudosuperior*	*A. pseudosuperior*	11	1	$p < 0.004$
24	*A. major*	*A. pseudosuperior*	4	3	$p = 0.706$
17	*A.* n.sp. nr. *reflexus*	*A.* n.sp. nr. *reflexus*	12	1	$p < 0.003$
23	*A. major*	*A.* n.sp. nr. *reflexus*	4	3	$p = 0.706$

We then shifted focus to species of the subgenus *Arrenurus*, testing four species: *Arrenurus americanus* Marshall, *Arrenurus major* Marshall, *A. pseudosuperior*, and *Arrenurus* n. sp., near *reflexus*. Males of *Arrenurus americanus* Marshall, *A. pseudosuperior*, and *Arrenurus* n. sp., near *reflexus* all yielded positive responses to water from females of their own species (Tables 4 and 5) but not to heterospecific sources (Table 4). Female *A. major* could not be reliably identified and were excluded, but males were tested with female-conditioned water from the other three species, arresting and orienting only to water from *A. americanus* females (Table 5).

The remainder of the study was concerned with testing whether pheromone responses occurred between species of different subgenera. Three species representing different subgenera were used in the fifth set of comparisons, namely

Table 6. Tests for pheromone response as indicated by male arrestant behaviour among species from the subgenera *Truncaturus*, *Megaluracarus*, and *Arrenurus* (see Table 2).

No. of males	Species of males	Female water	No. of males at test	No. of males at control	Sign test
40	*A. manubriator*	*A. manubriator*	30	0	$p < 0.001$
30	*A. rufopyriformis*	*A. rufopyriformis*	23	0	$p < 0.001$
19	*A.* n.sp. nr. *reflexus*	*A.* n.sp. nr. *reflexus*	10	1	$p < 0.006$
19	*A.* n.sp. nr. *reflexus*	*A. manubriator*	2	0	$p = 0.250$
40	*A. manubriator*	*A.* n.sp. nr. *reflexus*	1	0	$p = 0.500$
30	*A. rufopyriformis*	*A.* n.sp. nr. *reflexus*	0	0	$p = 1.000$
19	*A.* n.sp. nr. *reflexus*	*A. rufopyriformis*	1	0	$p = 0.500$

Table 7. Tests for pheromone response as indicated by male arrestant behaviour among species of the subgenus *Megaluracarus* and *Arrenurus* (see Table 2).

No. of males	Species of males	Female water	No. of males at test	No. of males at control	Sign test
40	*A. manubriator*	*A. manubriator*	25	0	$p < 0.001$
40	*A. americanus*	*A. americanus*	28	2	$p < 0.001$
40	*A. americanus*	*A. manubriator*	3	0	$p = 0.125$
40	*A. manubriator*	*A. americanus*	0	0	$p = 1.000$

Table 8. Tests for pheromone response as indicated by male arrestant behaviour among species of the subgenus *Megaluracarus* and *Arrenurus* (see Table 2).

No. of males	Species of males	Female water	No. of males at test	No. of males at control	Sign test
40	*A. manubriator*	*A. manubriator*	27	0	$p < 0.001$
10	*A. pseudosuperior*	*A. pseudosuperior*	7	0	$p < 0.008$
40	*A. manubriator*	*A. pseudosuperior*	0	0	$p = 1.000$
10	*A. pseudosuperior*	*A. manubriator*	0	0	$p = 1.000$

A. manubriator, *A. rufopyriformis*, and *A.* n. sp., near *reflexus*. In all three, female-conditioned water elicited a response by conspecific males but not by heterospecific males (Table 6). The same result occurred when *A. manubriator* was tested against *A. americanus* (Table 7) and *A. pseudosuperior* (Table 8).

Discussion

Female-emitted male-attractant sex pheromones appear to be widespread: we present evidence of chemical communication involving at least 10 species representing three subgenera of *Arrenurus*. In most cases, males reacted to conspecific female-conditioned water, but in *A. apetiolatus* we could only demonstrate that female-conditioned water elicited an arrestant response in male *A. manubriator*. All species demonstrating arrestant behaviour also exhibited the male readiness posture. Previous work by Proctor and Wilkinson (2001) showed that courtship and mating behaviour is similar in *A. manubriator*, *A. rufopyriformis*, and *A.* n. sp., near *reflexus*, three species also used in our study and representing three different subgenera. We suggest that female-emitted sex pheromones, the male arrestant response, male readiness posture, and the male's use of the fourth legs to grab the female are elements of a generalized courtship behaviour that is presumably plesiotypic for this genus. It is certainly not a universal pattern: for example, there are notable (and presumably apotypic) exceptions (e.g., *Arrenurus* (*Arrenurus*) *planus* Marshall; Proctor and Wilkinson 2001).

Unfortunately, it is difficult to interpret negative data: males typically will not arrest or exhibit a readiness posture and females will produce little if any pheromone unless they are well-fed and in good condition. When males of one species fail to respond to pheromones of another species, it may be because the pheromone truly fails to release arrestant behaviour or it could reflect poor condition of either the emmitters or receivers. Presumably, this is why *A. birgei* males would cluster and arrest near the source of female-conditioned water from *A. manubriator* but not to water from females of its own species, and conversely, male *A. manubriator* would respond to female-conditioned water from *A. apetiolatus* yet conspecific males were unresponsive. Subsequent to this study, we have found that laboratory-raised mites that have been segregated as males or females at adult transformation (and therefore, remain virgin) are the most responsive and reliable test organisms. Inconsistent results or lack of a pheromone response in the present study could be because of poor condition of the mites, or that in some species, previously mated females having reduced or curtailed their pheromone production. It is likely that most females used in our study were not virgins.

Apparently, it is quite common for members of the same species-group to arrest and orient to heterospecific pheromone sources. *Arrenurus manubriator*, *A. megalurus*, *A. marshallae*, *A. apetiolatus*, and *A. birgei* showed some degree of cross-species reactions, although not in all combinations. Male

A. manubriator and *A. megalurus* respond to female-conditioned water from each other's females, yet male *A. manubriator* but not *A. megalurus* responded to water from *A. apetiolatus* females. Several possible explanations exist: (1) species may use the same chemical or compound as a pheromone but have different threshholds (e.g., Kaae et al. 1973); (2) species may use different but related chemicals (in this case, *A. megalurus* would presumably have a pheromone intermediate in structure to chemicals used by the other two species); (3) pheromones consist of several chemicals in a blend, different species share some or all components, and the likelihood of interspecific responses depends on similarity of blends in constituent chemicals and ratios among those chemicals (typical with many insects, especially Lepidoptera, e.g., Löfstedt 1993).

Whereas there was cross-attractancy among many species within a subgenus, it did not extend across subgenera or throughout all members of a subgenus. Subgenera in the genus *Arrenurus* are largely considered to be artificial constructs, based upon superficial shape similarities of the adult males (Cook 1974). However, there are recognizable species groups (based on male morphology) within the subgenera that presumably reflect real phylogenetic affinities. Interspecific sex pheromone responsiveness provides additional insight, reinforcing concepts of species groups and implying relationships among these groups. Adult male *Arrenurus* (*Megaluracarus*) *megalurus*, *A. marshallae*, and *A. intermedius* are so similar morphologically that they intergrade into each other (*marshallae* group; Cook 1954b, 1976). Larvae of these species share a number of characters with *A. birgei* (*birgei* group; Cook 1954a, 1976) and *A. apetiolatus*, with a conspicuous synapomorphy being that seta Lh1 is located on coxal plate III rather than on the membranous idiosomal integument (Smith 1990). Larvae of *A. manubriator* are quite similar to those of *A. rotundus*; both have setae Lh1 in the plesiomorphic location on the idiosomal integument, but share with all of the forementioned species a broad differentiated marginal band on the dorsal plate and anterolateral corner of coxal plate III (Smith 1990), which again appear to be synapomorphies. These species appear to form a cohesive group within the subgenus *Megaluracarus*, shown above to have the same or similar sex pheromones. In contrast, *A. wardi* and *A.* n. sp., near *unisinuatus* do not show affinities with these species but have larvae very similar to those of a cluster of species characterized by a heart-shaped hump on the distal end of the male cauda (Cook 1954a); we call the *cardiacus* group. We did not see any evidence for pheromone communication within or between these two species, nor did they react to *A. manubriator* female-conditioned water. It is not clear whether *A. wardi* and *A.* n. sp., near *unisinuatus* belong to a clade of species that do not use female-emitted sex pheromones, or whether they were not producing and responding to pheromones because of suboptimal conditions in our experiments.

The subgenus *Arrenurus* includes a series of conspicuous species groups based upon male morphology (Cook 1974) but these are also reflected by similarities in larval form. The *americanus* group (Cook 1954b, 1976) includes *A. americanus* and *A. major*, the latter of which responded to female-conditioned water of the former but *A. americanus* did not respond to pheromone sources from other species of the subgenus. *Arrenurus pseudosuperior* belongs to the *superior* species group (Cook 1954b, 1976) and the one available male of *A. superior* responded to water from female *A. pseudosuperior* but *A. pseudosuperior* did not react to water conditioned by females of other species. *Arrenurus* n. sp., near *reflexus* was the only species in our study belonging to a third species group (we would cluster it with the *fissicornis* group; Cook 1954b) and only responded to water from its own females. Again, negative data are hard to interpret: there was no evidence of pheromone activity involving *A. magnicaudatus* (of the *superior* group) and both *A. fissicornis* and *A. fissicorniformis* (*fissicornis* group). Subsequent casual observations using laboratory-reared mites confirmed that *A. fissicornis* and *Arrenurus (Arrenurus) bleptopetiolatus* Cook (*fissicornis* group) males will respond to water conditioned by their own and each other's females.

The lack of species-specificity in chemical communication among many mites is a striking contrast to studies involving insects, as has previously been noted by Sonenshine (1985). This is not universal: Mori and Kuwahara (2000) found that various species of *Caloglyphus* spp. (Astigmata: Acaridae) did have distinctly different sex pheromones. It is controversial as to why there is species-specificity in sex pheromones for many insects, especially Lepidoptera. The classical paradigm is that it developed as a premating isolating mechanism to prevent hybridization, whereas another interpretation is that species-specific pheromones represent character displacement developed between already reproductively isolated species (Löfstedt, 1993). Regardless of why species-specificity is so widespread with insect sex pheromones, the question is why it is much less pronounced with mite sex pheromones.

There is strong potential for *Arrenurus* spp. males to be attracted to heterospecific females in their natural habitats. It is common to collect large numbers of *Arrenurus* spp. from one water body, and we have collected over 30 *Arrenurus* spp. from Telephone Bay of Lake Opinicon. It is not uncommon to get ten species from one net-full of substrate, and the species for which we demonstrated cross-attractancy commonly co-occur on this scale in various water bodies of Ontario. Heterospecific couplings between *Arrenurus* spp. regularly occur in containers when researchers are collecting and sorting living material; presumably this is an artifact of disturbance and artificially high densities, although perhaps such pairings occur naturally. One possibility is that sex pheromones in *Arrenurus* spp. are a long-range attractant, but that individuals can then use short-range species-specific cues or courtship behaviour to discriminate among potential mates. Another possibility is that stimulated males produce their own sex pheromone, which are species-specific. *Arrenurus* species with interspecific sex pheromone responsiveness usually have males that are unambiguously different in physical form, while congeneric species of insect that do not exhibit cross-attractancy are often difficult to identify using morphological characters (e.g., *Choristoneura* spp., Silk and Kuenen 1988; *Yponomeuta* spp., Löfstedt 1993). Males of many *Arrenurus* spp.

assume the readiness position and use their fourth legs to grab and grapple the female into position (Proctor and Wilkinson 2001), but the extravagance of male form and diversity among species (while females are cryptically similar; Cook 1976) strongly implies sexual selection and it is probable that female choice still operates to some degree despite male adaptations to circumvent it.

Acknowledgements

Part of this study was conducted at the Queen's University Biological Station and we appreciate the logistical support afforded us. Thanks also to Katherine Muma for helping with field collections, proofreading, and her comments on the manuscript.

References

Böttger K. 1962. Zur Biologie und Ethologie der einheimischen Wassermilben *Arrenurus* (*Megaluracarus*) *globator* (Müll.), 1776 *Piona nodata nodata* (Mll.), 1776, und *Eylais infundibulifera meridionalis* (Thon), 1899 (Hydrachnellae, Acari). Zool. Jb. Syst. 89: 501–584.

Cook D.R. 1954a. Preliminary list of the Arrenuri of Michigan. Part II. The subgenus *Megaluracarus*. Trans. Am. Microsc. Soc. 73: 367–380.

Cook D. 1954b. Preliminary list of the Arrenuri of Michigan. Part I. The subgenus *Arrenurus*. Trans. Am. Microsc. Soc. 73: 39–58.

Cook D.R. 1974. Water mite genera and subgenera. Mem. Am. Ent. Inst. 21: 1–860.

Cook D.R. 1976. North American species of the genus *Arrenurus*, mostly from Florida (Acarina: Arrenuridae). Contrib. Am. Ent. Instit. 11: 1–58.

Kaae R.S., Shorey H.H. and Gaston L.K. 1973. Pheromone concentration as a mechanism for reproductive isolation between two lepidopterous species. Science 179: 487–488.

Löfstedt C. 1993. Moth pheromone genetics and evolution. Philos. Trans. Biol. Sci. 340: 167–177.

Mori N. and Kuwahara Y. 2000. Comparative studies of the ability of males to discriminate between sexes in *Calloglyphus* spp. J. Chem. Ecol. 26: 1299–1309.

Proctor H.C. and Smith B.P. 1994. Mating behaviour of the water mite *Arrenurus manubriator* Marshall (Acari: Arrenuridae). J. Zool. Lond. 232: 473–483.

Proctor H.C. and Wilkinson K. 2001. Coercion and deceit: water mites (Acari: Hydracarina) and the study of intersexual conflict. In: Halliday R.B., Walter D.E., Proctor H.C., Norton R.A. and Colloff M.J. (eds), Acarology: Proceedings of the 10th International Congress. CSIRO Publishing, Melbourne, pp. 155–169.

Roelofs W.L. 1995. Chemistry of sex attraction. Proc. Natl. Acad. Sci. 92: 44–49.

Silk P.J. and Kuenen L.P.S. 1988. Sex pheromones and behavioral biology of the coniferophagous *Choristoneura*. Ann. Rev. Entomol. 33: 83–101.

Smith B.P. 1990. Descriptions of larval *Arrenurus bartonensis*, *Arrenurus birgei*, *Arrenurus neobirgei*, and *Arrenurus rotundus* (Acari: Hydrachnidia; Arrenuridae). Can. Entomol. 122: 77–91.

Smith B.P. 1998. Loss of larval parasitism in parasitengonine mites. Exp. Appl. Acarol. 22: 187–199.

Smith B.P. and Hagman J. 2002. Experimental evidence for a female sex pheromone in *Arrenurus manubriator* (Acari: Hydrachnida; Arrenuridae). Exp. Appl. Acarol. 27: 257–263.

Smith I.M., Cook D.R. and Smith B.P. 2001. Water mites (Hydrachnida) and other arachnids. In: Thorp J.H. and Covich A.P. (eds.), Ecology and Classification of North American Freshwater Invertebrates, 2nd ed. Academic Press, pp. 551–659.

Sonenshine D. 1985. Pheromones and other semiochemicals of the Acari. Ann. Rev. Entomol. 30: 1–28.
Viets K.O. 1987. Die Milben des Süsswassers (Hydrachnellae und Halacaridae [part.], Acari). 2: Katalog. Sonderbände des Naturwissenschaftlichen Vereins in Hamburg 8. Verlag Paul Parey, Hamburg and Berlin, pp. 1012.

Experimental and Applied Acarology **34:** 127–147, 2004.

Red, distasteful water mites: did fish make them that way?

HEATHER C. PROCTOR[1,*] and NEERA GARGA[2]
[1]*Department of Biological Sciences, University of Alberta, Edmonton, Alberta, Canada T6G 2E9;* [2]*3-2636 First Avenue, NW, Calgary, Alberta, Canada T2N 0C9; *Author for correspondence (e-mail: hproctor@ualberta.ca; phone: +1-780-492-5704; fax: +1-780-492-9234)*

Key words: Anti-predator defenses, Assemblage structure, Fish, Müllerian mimicry, Photoprotection

Abstract. Water mites (Acari: Hydrachnida) are unusual among the typically cryptic freshwater fauna in that many species are brightly colored red or orange, and also appear to be distasteful to fish. This apparent aposematism (use of color to warn predators) has been previously explained as the evolutionary end-product of pressure from fish predation. The fish-predation argument has been supported by observations that fish spit out red mites, powder made from red water mites is more distasteful to fish than powder made from non-red mites, and red mites appear to be more abundant than non-red mites in water bodies where fish are present. In this paper, we challenge the hypothesis that fish were the sole driving force behind the evolution of aposematism in water mites. We show that non-red mites actually dominate in water bodies with fish, and that red mites are more abundant in temporary, fishless water bodies. We also demonstrate that powder made from red, terrestrial velvet mites (Trombidiidae) was as distasteful to fish as powder made from red water mites. We suggest that the main role of red and orange carotenoid pigments may be to act as photoprotectants, and hypothesize that redness originated in the terrestrial ancestors of water mites and has been retained in certain lineages of water mites after the invasion of the aquatic habitat. We also suggest that distastefulness evolved subsequent to bright coloration in response to increased conspicuousness to predators. Relaxed selection for redness has occurred when adults and/or larvae are less exposed to sunlight, either through occupying more protected habitats, parasitizing more nocturnal hosts, or parasitizing hosts for a short period of time. Our ability to test this alternative hypothesis is hampered by lack of knowledge of the source and mode of action of distastefulness, and of phylogenetic relationships among the Parasitengona.

Introduction

Many arthropods use conspicuous signals, such as bright colors or striking patterns, to warn predators that they are foul-tasting or dangerous. This combination of high visibility and unprofitability is termed 'aposematism' (apo – away from, sema – sign) (Lindström 1999; Joron 2003). Although most researchers would agree that the pairing of conspicuousness with distastefulness is an effective anti-predator mechanism, there is no consensus on how the two characteristics come to co-occur over evolutionary time (Lindström 1999; Merilaita and Kaitala 2002; Joron 2003). One important question is how aposematism can spread in a population if the first, rare, brightly colored mutants are more likely to be seen by a predator (e.g., as in passerine birds, Huhta et al. 2003). One of the first hypotheses proposed to explain the evolution of aposematism was kin selection (Turner 1971; Leimar et al. 1986; Guilford and Cuthill 1991). Genetically related individuals that share bright colors and unpalatability would suffer from predation on an individual level, but each death would serve to deter that particular predator from eating the martyr's kin. This effect is greatly enhanced if kin aggregate, as occurs in many juvenile insects such as caterpillars and sawfly larvae (Joron 2003; but see Tullberg et al. 2000). However, this hypothesis cannot explain the evolution of aposematism in organisms where kin are widely dispersed (e.g. among animals with planktonic larvae, or where dispersal occurs prior to development of aposematic signals). There one must turn to individual selection. For individual selection to explain the origin of aposematism, two features must be present (Joron 2003). First, distastefulness must be sufficiently strong and immediate to cause the predator to reject the prey immediately rather than after digesting it. Second, prey must possess traits that increase their chance of surviving an encounter with a predator, such as a tough integument or flexible appendages.

Both kin- and individual-selection approaches typically assume that distastefulness evolves first, usually as a byproduct of diet (e.g. sequestration of toxic plant chemicals), and warning coloration evolves thereafter. A seldom-considered alternative hypothesis is that conspicuous coloration may evolve as a result of some other selection pressure (sexual selection, thermoregulation, etc.), and only in unpalatable prey can brightness be maintained in the face of predator pressure (Guilford and Dawkins 1993; Summers and Clough 2001). Both hypotheses are intrinsically historical (i.e., character A evolutionarily precedes character B) and could be tested by mapping the evolution of characters onto a phylogeny; however, there are very few tests that use a phylogenetic approach. Summers and Clough (2001) found that brightness and toxicity evolved in synchrony in dendrobatid frogs, but their cladogram could not resolve which feature came first. Sillén-Tullberg (1988) and Tullberg and Hunter (1996) found that aposematism preceded gregariousness of kin in several lineages of Lepidoptera, but could not disentangle the relative times of origin of bright color and anti-predator defense.

As well as biases in the mode of hypothesis testing, studies of aposematism almost invariably deal with terrestrial and marine systems, where brightly colored organisms are common and span a wide taxonomic range (Pearse et al. 1987; Ang and Newman 1998). In sharp contrast, freshwater systems have a paucity of brightly colored animals (Pennak 1976; Kerfoot 1982; Pearse et al. 1987). The most obvious exceptions to the drab inhabitants of fresh water are the water mites (Acari: Parasitengona; Hydrachnida, Hydracarina, Hydrachnidia or Hydrachnellae, depending on the nationality of the author). These mites often sport brilliant colors such as scarlet and orange as well as more subdued hues of blue, purple and green. The unusual brightness of water mite coloration has long been of interest to naturalists and ecologists (Dalyell 1851;

Elton 1923; Kerfoot 1982). Elton (1923) was particularly impressed by scarlet species and wondered whether these mites used color to advertise distastefulness. He found that starved three-spined stickleback fish (*Gasterosteus aculeatus* L.) would spit out red water mites, leaving the mites apparently unharmed. Elton hypothesized that the presence of red coloration in several different water mite genera was an example of convergence in aposematic patterns, i.e. Müllerian mimicry. Many years later, Kerfoot (1982) continued the investigation into water mite aposematism. He determined that flour balls including powder from brightly colored water mites were more distasteful to guppies (*Poecilia reticulata* Peters) than were balls including powder from drably colored species. He attributed the source of distastefulness to phenolic secretions from the mites' skin glands (glandularia). Kerfoot (1982) also compared assemblages of mites in eight ponds in New Hampshire, and found that ponds with fish tended to have more individuals of brightly colored species. He concurred with Elton's (1923) Müllerian mimicry argument that red, distasteful water mites represent '...convergence of many unrelated lineages...' in response to fish predation (Kerfoot (1982), p. 546).

Despite the apparent strength of Kerfoot's (1982) hypothesis, there are some difficulties with his conclusions. First, more than 15 years of field collections by the senior author, spanning the width of North America and all latitudes of Australia, suggest that red mites are most abundant in temporary water bodies, which lack fish. Second, the Müllerian mimicry hypothesis that redness evolved repeatedly within water mites does not take their phylogeny into account. The Hydrachnida are likely derived from terrestrial ancestors in the prostigmatan cohort Parasitengona (Krantz 1978; Welbourn 1991). The two major terrestrial lineages, Erythraeina and Trombidiina (Welbourn 1991) are made up mainly of large, soft-bodied, red or orange mites, characteristics similar to those of many water mites. These similarities suggest that redness may not be a novel trait that evolved when terrestrial mites invaded their new aquatic habitat, but is rather a symplesiomorphy that first arose on land. Evidence in favour of this interpretation is that taxa composed primarily of red species (e.g. Eylaoidea, Hydryphantoidea) are thought to be more primitive than taxa displaying a range of colors (e.g. Arrenurioidea, Hygrobatoidea) (based on cladogram in Harvey 1998). An alternative hypothesis of water mite origins places them at the base of the Parasitengona with terrestrial members of the Anystae as immediate ancestors (Witte 1991; Smith et al. 2001). Because Anystae are also mainly red and orange in color (e.g. the whirligig mites, Anystidae), this does not substantially alter the scenario of redness being ancestral in the Hydrachnida. Elton (1923) recognized that many land mites are bright scarlet but did not test whether they were distasteful. A recent review of the ecology of the Trombidiidae quoted only anecdotal evidence that velvet mites are distasteful to predators (Zhang 1998). Thus it is currently unclear whether water mite ancestors were aposematic, or simply brilliantly colored.

Here we challenge the assumption that red mites are associated with fish, either ecologically or evolutionarily. By sampling a greater number and a wider range of water bodies than Kerfoot (1982), and using a more objective approach to color classification, we more thoroughly test the hypothesized correlation between presence of fish and high proportion of red-colored mites. We also determine whether a red terrestrial relative of water mites is as distasteful to fish as are red water mites. Finally, we determine the probability of water mites surviving attack by fish, a necessary outcome if aposematism is to be maintained by individual rather than kin selection.

Materials and methods

Study sites

Water mites were collected from 47 sites in eastern Ontario, Canada, in spring and summer of 1995. From May 10 to May 25, mites were collected from 11 water bodies situated around the Wildlife Research Station in northern Algonquin Provincial Park (46°36′N, 78°31′W). Six of these sites had fish. In April, July and August they were collected from the remaining 36 sites in the area surrounding the Queen's University Biological Station at Chaffey's Locks, Leeds County (44°34′N, 79°15′W). Twenty-one of these sites had fish. We determined whether fish were present or absent at each site using minnow traps, *in situ* observations, and historical records (word of mouth) of the area. The types of water bodies sampled included temporary vernal ponds and marshes, permanent ponds, marshes, lakes, and a few slow-flowing streams and rivers. All fishless sites were temporary water bodies. For a full list of sites, see Appendix 1 in Garga (1996).

Water mite collection and identification

Water mites were collected using a 500 μm dip-net and sieves. The dip-net was used to sweep through vegetation and stir up substrate. At each site as many sweeps were carried out as were needed to fill a collection container ~750 ml in volume. Sampling took place in the littoral zone of all sites from the shoreline to a depth of approximately 1.5 m. The collected material – composed of mud, leaves, water plants and other debris – was poured through a 5 mm sieve onto a collecting sieve of 500 μm mesh. Plants and detritus were rinsed in the 5 mm sieve with water from the sampled site until no additional material passed through to the smaller mesh. Sorting of water mites was done in the laboratory or at the field site if it was sufficiently bright. The concentrated collected material was placed into the centre of a white enamel dissecting tray filled with approximately 5 cm of water. Water mites were removed from the tray using a pipette until several minutes had passed during which no mites were seen moving. Mites were keyed to genera using Smith (1990) while they were still alive. Prior to measuring the color of the water mites (see below), we grouped all mites of the same genus and of similar coloration together in a single well of a six-well tissue-culture plate. We used genus as the finest taxonomic level because this allowed us to categorize deutonymphs as well as adults (deutonymphs can rarely be identified to species-level with confidence). Genera in the hydryphantid subfamily Thyasinae and the aturid subfamily Axonopsinae could not readily be distinguished based on live specimens, and so were identified only to subfamily.

Color quantification

To avoid subjective variation in color matching, the following procedure was undertaken by a single person (NG). Water mite color was quantified using a two-step process. One live water mite from each genus/color group was placed in a clear tissue-culture well-plate under a dissecting microscope. The body color of the water mite was then matched to a Canadian Tire Corporation paint chip from one of 1200 possible paint chips. Each paint chip had a unique name and numerical code. After the best color match for the mite was found, the name and number of that paint chip was assigned to all individuals in the well that initially held the measured individual. In the second step of color quantification, the selected paint chips were analyzed using a colorimeter (Gretag Macbeth Eye^{TM} spectrophotometer) in the Department of Art Conservation at Queen's University, Kingston, Ontario. The colorimeter broke each color into three components each ranked from 1 to 11: the degree of redness on a scale of green–red, the degree of yellowness on a scale of blue–yellow, and degree of brightness on a scale of dark to light (Figure 1). The color reading for each paint chip was assigned an x–y coordinate value based on the red–green axis and the blue–yellow axis. We selected the paint chips that we considered 'red', and then determined where they fell on the color axes. We found that 'red' colors scored 8 or more on the red scale (x-coordinate) and 8 or more on the yellow scale (y-coordinate). This enabled us to determine what proportion of water mites from each site was colored red.

Although it is not necessary for testing previous assertions by humans that red mites are more abundant in water bodies with fish, a sensible question is whether our assessment of 'red' was likely to match what fish see. The fish species found in the water bodies we sampled can be classified as generalized diurnal mid-water species (Douglas and Djamgoz 1990). Typically, these fish have: (a) blue-sensitive single cones maximally sensitive (λ_{max}) at 460 nm, and (b) double cones than contain green-sensitive pigments with λ_{max} at 540 nm and a longer wave, red-sensitive pigment with λ_{max} in the 580–620 nm range. Primate color vision is the result of pigments with maximal sensitivities at 445 nm (blue), 530 nm (green), and 625 nm (red) (Eckert et al. 1988). While fish only have two kinds of visual pigments, the existence of the double cone makes their vision similar to the three-pigment system of humans. One species of freshwater fish found in many of our fish-present sites, *Lepomis macrochirus*

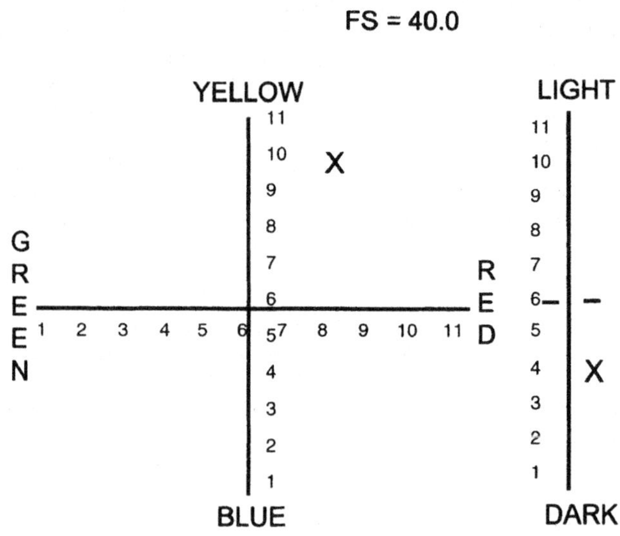

Figure 1. Sample printout of color analysis from MacBeth Eye. The green–red, blue–yellow and light–dark axes were all scored from 0 to 11. This sample color scored 8/10/4 (red/yellow/dark).

Rafinesque (bluegill sunfish), has few blue-sensitive cones and therefore has little short-wavelength sensitivity (Douglas and Djamgoz 1990). Because it is short wavelength light that is most altered by scatter and chromatic attenuation of light, bluegills are relatively unaffected by these confounding phenomena and view prey items with minimal color aberration. This aspect of bluegill color vision, and the fact that they are a common species in the sampled area, made them a good predator to use with taste test and survival experiments with water mites (see below). Both bluegill and pumpkinseed sunfish, *Lepomis gibbosus* (L.), show intensified orange and red coloration in breeding males (Scott and Crossman 1973); this is good evidence that these colors are used for intraspecific visual signaling, and that the sensitivity of the fishes' retinas to these wavelengths is not simply a laboratory artifact.

Fish-predation studies

(a) Distastefulness of terrestrial relatives of water mites

Following the flour-ball protocol developed by Kerfoot (1982), we tested whether fish found red terrestrial mites (*Dinothrombium magnificum* (Le Conte), Parasitengona: Trombidiidae) to be distasteful. Mites were collected in Arizona in July 1995 by B. Tomberlin (Hatari Invertebrates, Portal, Arizona). *Dinothrombidium magnificum* are large, soft bodied and bright red. Our specimens had an average length of 1.5 cm and a width of 0.5 cm. These mites are reported to be rejected by ants, lizards and toads, and to taste unpleasant to humans (pers. comm. 2003, J. Schmidt, Carl Hayden Bee Research Center, Tucson, Arizona). We used wild-caught bluegill sunfish (*L. macrochirus*) from Lake Opinicon as predators. Experiments took place from September 1995 to February 1996 at Queen's University, Kingston, Ontario. Fish were contained in a 100-gallon tank (ca. 400 l) with clear partitions to separate the individual fish during the testing trial. A blind was erected in front of the testing aquaria because we noted that fish would alter their feeding behavior if they saw humans watching them. Fish were maintained on a diet of Tetra-Min fish flakes, and were starved 24 h prior to being tested.

Mites were air-dried and ground into a powder using a mortar and pestle. The powder was mixed in different proportions, by weight, with commercial white flour. We made small pellets ('flour balls') by moistening the mixture and rolling it into spheres ~2 mm in diameter. At the start of every feeding trial, fish received control flour balls that did not contain any mite powder. If a fish rejected the control balls, it was not included in the experiment. In any given trial, a single test ball made with mite powder was presented to a single fish and we noted the number of times the fish mouthed the test ball, and whether the fish accepted or rejected the test ball. The dilution series included the following concentrations (% mite by mass): 33, 25, 16, 6, 3% and each fish was tested once at each concentration. All fish were tested at one concentration of mite powder before moving on to the next concentration in order of decreasing strength (the method followed by Kerfoot 1982). Five days separated the rounds of testing at one concentration. Sample size of fish was initially 23, but mortality reduced this to 20 by the end of the *D. magnificum* experiment, and to 17 soon thereafter.

The remaining 17 fish were presented with balls containing 33.3% powder made from *Limnochares americana* Lundblad (Limnocharidae). Kerfoot (1982) found that flour balls made with powder from a *Limnochares* sp. were rejected at a low percentage of powder by guppies (Poeciliidae: *Lebistes reticulatus*).

(b) Survivorship of water mites after mouthing by fish

Species from four genera of water mites (*Lebertia*, *Hydrodroma*, *Limnochares*, *Limnesia*) were tested for their ability to survive a predator's attack. The predators were two species of freshwater fish commonly encountered in the bodies from which the water mites were removed: *Lepomis gibbosus* (L.) (pumpkinseed sunfish) and *L. macrochirus* (bluegill sunfish). These experiments were carried out at the Queen's University Biological Station from June 1995 to August 1995. The fish were removed from the holding tank, placed into testing aquaria, and starved 24 h prior to being tested. Using a pipette, one of us (NG) introduced a live water mite into the aquarium and moved behind a blind to observe the number of times the fish mouthed the mite and whether the mite was eaten. After an attack occurred, if the mite was still alive, it was removed from the aquarium and placed into a holding well. The mite's condition was monitored for the following 48 h after which it was scored as alive or dead.

Pigments in water mites

In order to interpret function of color in water mites and their relatives, it was necessary to know what these pigments are. Therefore, we performed a literature search for pigments found in water mites and terrestrial parasitengones.

Statistical analyses

A two-by-two goodness-of-fit test was performed using the software package Statistix 3.1 (N.H. Analytical Software 1989) to determine if a greater proportion of red water mites than expected was collected from fish-absent sites than from fish-present sites. All other statistics were calculated using SigmaStat 2.03 (Fox et al. 1995). Because data from the colorimetry analysis were not normally distributed (Zar 1996), non-parametric Wilcoxon rank sum tests were performed to determine if any of the three aspects of water mite coloration (red–green, yellow–blue, dark–light) differed between fish-absent and fish-present bodies of water.

Results

Red water mites in fish-absent and fish-present water bodies

In total, 2637 water mites were collected from 27 fish-present bodies of water, and 4049 from 20 fish-absent sites. All fish-absent sites were also temporary pools or marshes. Twenty-three genera of water mites were represented, as well as unidentifiable Thyasinae and Axonopsinae (Table 1). Of the 6686 mites examined, 4861 individuals were categorized as 'red' (72.7% of all individuals). In fish-absent water bodies, a significantly greater proportion of mites were red (89.7%), while in fish-present sites only 46.5% were red (Yate's corrected $\chi^2 = 1496.7$, $p < 0.0001$). Figure 2 shows the percentage of red mites at each site type, including only those sites where 10 or more mites were collected. After arcsine-square root transformation, these data were compared using a one-way ANOVA. Fish-absent sites ($N = 15$) had a significantly greater mean proportion of red mites than fish-present sites ($N = 25$) (df $= 39$, $F = 37.3$, $p < 0.0001$). For each of the four most common genera of water mites that also showed variation in body color (*Arrenurus*, *Limnesia*, *Piona* and *Tiphys*), there was also a significantly higher proportion of red individuals in fish-absent water bodies (Figure 3) (Yate's corrected χ^2 range 54–638, all p's < 0.0001).

Color component analysis for fish-absent and fish-present bodies of water

We used 84 of the 1200 potential paint chips to assign colors to water mites. Fewer colors were represented in mites from fish-absent water bodies than in mites from fish-present ones (Table 2). The mite assemblages of fish-absent

Table 1. Water mites collected in this study categorized by percentage of individuals that were classified as 'red'.

> 96% of individuals red	30–89% of individuals red	< 15% of individuals red	Individuals never red
Atractides (Hygrobatidae)[a] (138:3)	*Arrenurus* (Arrenuridae) (365:773)	*Mideopsis* (Mideopsidae) (6:51)	*Albia* (Aturidae) (0:16)
Bandakia (Anisitsiellidae) (1:0)	*Frontipoda* (Oxidae) (20:23)	*Neumania* (Unionicolidae) (14:140)	Axonopsinae (Aturidae) (0:16)
Eylais (Eylaidae) (9:0)	*Hygrobates* (Hygrobatidae) (3:1)	*Oxus* (Oxidae) (2:17)	*Koenikea* (Unionicolidae) (0:13)
Forelia (Pionidae) (7:0)	*Lebertia* (Lebertiidae) (16:33)		*Unionicola* (Unionicolidae) (0:203)
Hydrachna (Hydrachnidae) (31:1)	*Limnesia* (Limnesiidae) (369:331)		*Wettina* (Wettinidae) (0:3)
Hydrodroma (Hydrodromidae) (527:1)	*Piona* (Pionidae) (654: 80)		
Hydryphantes (Hydryphantidae) (181:2)	*Tiphys* (Pionidae) (147:94)		
Limonochares (Limnocharidae) (38:0)			
Sperchon (Sperchontidae) (2:0)			
Thyasinae (Hydryphantidae) (2355:0)			

Numbers in parentheses are number red:number non-red.
[a]Familial classification follows Smith et al. (2001).

water bodies had a greater representation of 'red' paint-chip colors than did those from fish-present water bodies (Mann–Whitney rank sum test, $t = 3797$, $p < 0.0001$) (Table 2). A significant difference between fish-present and fish-absent sites also existed along the blue–yellow axis, with mite colors from fish-absent water bodies having more yellow (Mann–Whitney rank sum test, $t = 3513$, $p = 0.004$). There was no significant difference between site types in the dark–light axis for water mite coloration (Mann–Whitney rank sum test, $t = 2591$, $p = 0.076$).

Distastefulness of terrestrial velvet mites to sunfish

In dried form, the red terrestrial mite *D. magnificum* appeared to be distasteful to bluegill sunfish (Figure 4). At 16% and higher concentration of terrestrial

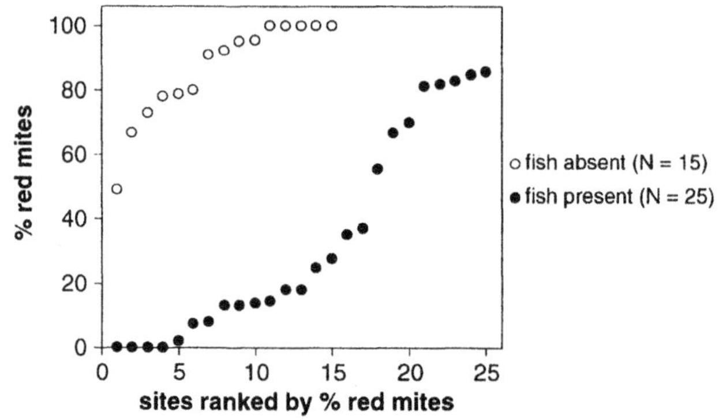

Figure 2. Percentage of red water mites at fish-absent and fish-present sites. Sites are ranked from lowest to highest percentage red within each site type. Only those sites with 10 or more mites are shown ($N = 10$ from Algonquin Provincial Park, $N = 30$ from Queen's University Biological Station). Symbols for sites that tied at 0% red are spread out along the *x*-axis rather than being overlapped.

mite powder, flour balls were rejected >90% of the time. Powder from *D. magnificum* appeared to be more distasteful to fish than powder from the red water mite *L. americana*, as flour balls made with 33% powder from the latter mite were rejected only about 50% of the time (Figure 4, square symbol). In contrast, Kerfoot (1982) found that guppies rejected flour balls containing 10% powder from an unnamed *Limnochares* species 100% of the time. It is not clear whether this difference in palatability of *Limnochares*-powder balls is due to our having used a different fish species, a different type of flour, or different species of *Limnochares*.

Mortality of water mites after attacks by L. gibbosus *and* L. machrochirus

All 51 water mites presented to fish were mouthed at least once. Twelve mites were eaten, and 39 mites were released (Figure 5). Of these, 38 survived mouthing and one died. One *Lebertia* sp. survived being mouthed and released nine times. None of the *Limnochares americana* were eaten, which suggests that they were the most distasteful of the four species presented.

Pigments in water mites

Several researchers have examined water mite pigments using column- and thin-layer chromatography (Green 1964; Czeczuga and Czerpak 1968a, b, c; Meyer and Kabbe 1991). Most of the mites examined were red-colored species that were found to contain carotenoid pigments (Table 3). Meyer and Kabbe (1991) tested one non-red water mite species, *Hydrodroma danuveinsis*, which

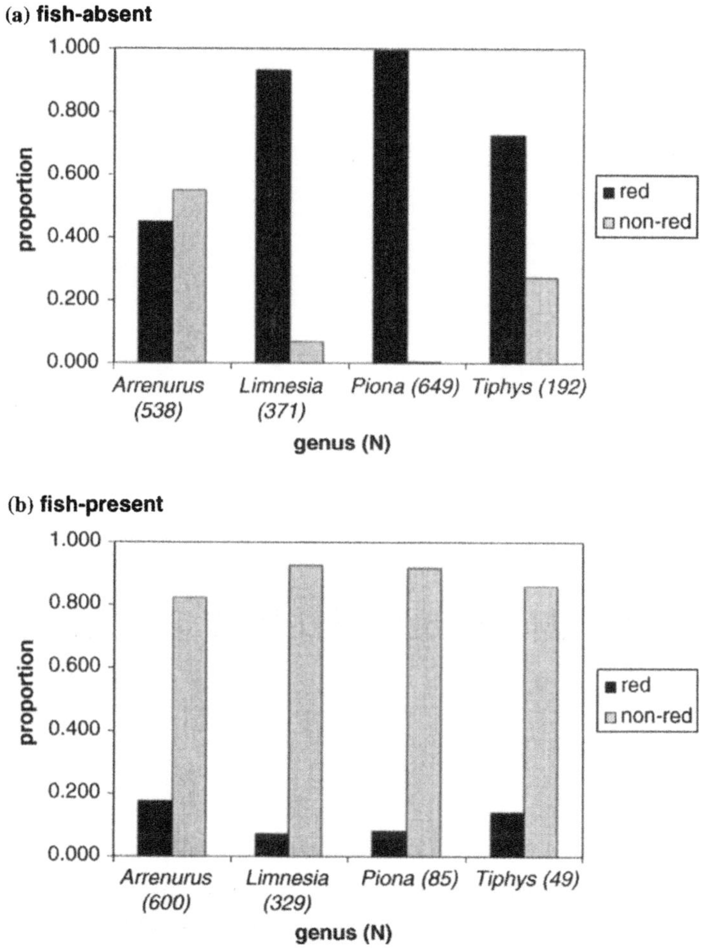

Figure 3. Proportions of red and non-red individuals of four of the most abundant mite genera: (a) in water bodies that lacked fish, (b) in water bodies with fish.

Table 2. Differences in fish-absent and fish-present sites with respect to mean values of the three components of measured coloration of water mites (see also Figure 1 for an illustration of the components).

	Green (1)–red (11)	Blue (1)–yellow (11)	Dark (1)–light (11)
Fish-absent sites (N = 47 colors)[a]	9.87	8.39	1.15
Fish-present sites (N = 82 colors)	7.62	7.64	1.74
Wilcoxon rank sum W statistic	3769	3513	2590
p	< 0.0001	0.0043	0.0759

[a]Sample size refers to the number of different colors observed among water mites at each habitat type.

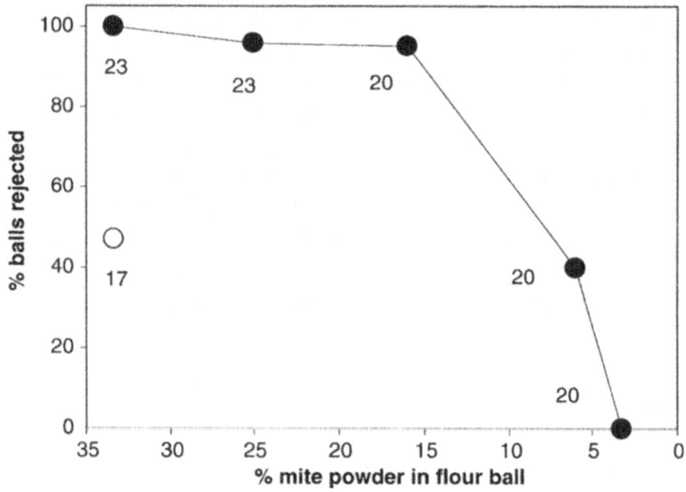

Figure 4. Percentage of flour balls rejected by *Lepomis macrochirus* at different proportions of velvet mite powder (closed symbols). Numbers in parentheses represent numbers of trials with different fish where one trial = one ball presented to one fish. Single open symbol shows rejection rate for flour balls made with *Limnochares americana* powder.

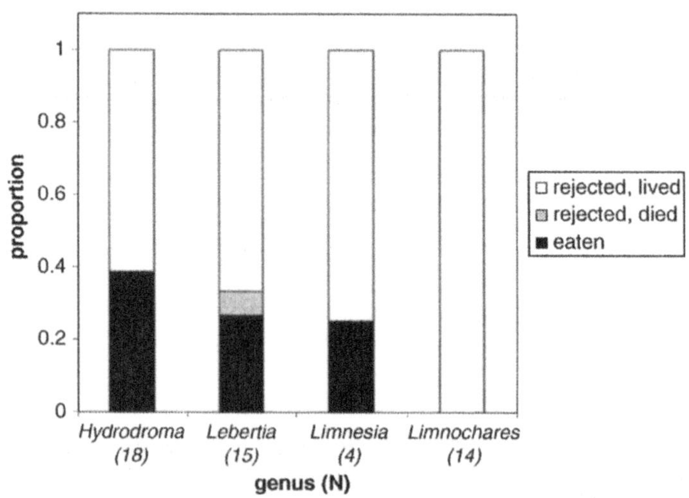

Figure 5. Survivorship of water mites when presented to sunfish (*Lepomis macrochirus, L. gibbosus*).

did not contain carotenoids. Carotenoids are also responsible for bright red coloration of some terrestrial parasitengones. Metcalf and Newell (1962) found that an *Erythraeus* sp. (Erythraeidae) contained pigments similar to astaxanthin, which is found in some red water mites (Table 3). Red velvet mites in the genus *Trombidium* (Trombidiidae) also contain carotenoids (Manunta 1939).

Table 3. Summary of carotenoids extracted from water mites (Numbers in parentheses indicate references).

Carotenoid	*Hydrachna geographica* (1)	*Piona nodata* (1)	*Hydryphantes dispar* (2)	*Eylais extendens* (3)	*Limnochares aquatica* (4)	*Hydrodroma despiciens* (4)	*Eylais hamata* (5)	*Hydrodroma danuveinsis*[a] (4)
Astacene	✓	✓		✓			✓	
Astaxanthin			✓				✓	
β-Carotene	✓	✓	✓		✓		✓	
Canthaxanthin			✓		✓		✓	
Cryptoxanthin	✓	✓						
Diepoxide					✓			
Echinenone	✓	✓	✓		✓	✓	✓	
Keto-carotenoid				✓				
Lutein	✓	✓	✓	✓	✓	✓	✓	
Neochrone					✓			
Neoxanthin			✓					
Torulen						✓		
Violaxanthin						✓		

[a]A yellow–brown species (1) Czeczuga and Czerpak (1968a), (2) Czeczuga and Czerpak (1968b), (3) Green (1964), (4) Meyer and Kabbe (1991), (5) Czeczuga and Czerpak (1968c).

Discussion

No correlation between fish presence and dominance of red mites

Our findings on water mite color-assemblages did not support Kerfoot's (1982) observations that fish presence is correlated with a high proportion of red mites. In fact, they show the opposite, that red water mites are more common in water bodies that lack fish. One explanation for the difference between the two sets of results might be the small number ($N = 8$) and close proximity of water bodies sampled by Kerfoot (1982). In contrast, we collected from many sites ($N = 47$) that differed in permanence, substrate, clarity of water, depth, and size. Within our sample, there were some fish-absent and fish-present sites that opposed the general trend but they each represented a small percentage of the total number of sites (5 and 19%, respectively) (complete data in Garga 1996). Additional support for our observation of more red mites in fishless sites is that within genera that contained both red and non-red individuals, red individuals also dominated in fish-absent water bodies (Figure 3). One caveat in making a direct comparison between Kerfoot's (1982) findings and ours is that our fish-absent sites were temporary water bodies, whereas Kerfoot's may have been permanent; however, this is not clearly stated in the paper.

Water mites can survive fish attack

Except for those that were eaten, almost all of the mites survived being mouthed (usually repeatedly) by fish (Figure 5). Thus, these water mite species meet Joron's (2003) two criteria for having evolved aposematism through individual selection: having a defense that acts rapidly, and being able to survive the process of being 'sampled' by a predator. Given the life cycle of most water mites, one would have rejected the potential for a kin-selection origin of aposematism in this group regardless of the outcome of our experiment. With few exceptions, water mite disperse from their natal area as parasites of flying insects (Smith et al. 2001), and so it is very unlikely that related nymphs and adults would co-occur at the fine spatial scale needed for kin-selection to function. However, as we argue below, although predation by fish may be one factor maintaining distastefulness in water mites, it does not appear to be the only force behind the bright coloration of many mite species.

Pigmentation as protection from damaging light

Our field data suggest that fish predation cannot be the sole factor maintaining red pigmentation in water mites. Redness in water mites is caused primarily by carotenoids (Table 3), which are also present in terrestrial parasitengones. One well-known function of carotenoid pigments is their ability to prevent light-induced oxidation of sensitive molecules (Hairston 1979a, b). Red pigmented crustaceans are more common in water bodies that experience intense solar radiation such as shallow ponds or alpine lakes (Hairston 1981; Luecke and O'Brien 1981). Copepods with high levels of carotenoids such as astaxanthin can survive exposure to more intense radiant energy than can paler copepods (Hairston 1979a; Ringelberg et al. 1984; Chalker-Scott 1995). Carotenoid-bearing copepods are also more likely to swim near the surface of a water body than are pale copepods (Hairston 1980). Redness does not appear to act as a deterrent to predation *per se*, as highly pigmented zooplankton are more readily seen and eaten by fish than are paler ones (e.g. Hairston 1981; Luecke and O'Brien 1983). Likewise, redness in water mites may not act as an *a priori* deterrent; Kerfoot (1982) found that fish did not show an innate aversion to red-colored flour balls, and did not learn to avoid eating red balls laced with mite powder any faster than black balls.

All of the fish-absent sites in our survey were shallow, temporary waterbodies. This corresponds well with the type of freshwater habitat typically occupied by red-pigmented copepods. Thus it may be that dominance of these fishless sites by red water mites was also influenced by the greater penetration of damaging radiant energy. Unlike with copepods, however, there have been no experimental studies demonstrating that the carotenoids in red water mites protect them from UV and other damaging radiation. Meyer and Kabbe (1991) also considered the possibility that carotenoids could serve as UV protectants for red water mites of the genera *Limnochares* and *Hydrodroma*. They noted, however, that *Hydrodroma* tend to keep themselves well hidden among water plants and mud, and hence are seldom directly exposed to sunlight. We observed a similar conflict in our own data, because many of the temporary ponds were shaded by trees and had a thick layer of submerged leaves and mud in which mites were often hidden.

However, adult water mites may accumulate carotenoids not for their own photoprotection, but for that of their offspring. Animals cannot produce carotenoids themselves and must garner them by consuming plants or herbivorous animals (Goodwin 1984). These pigments can be deposited in eggs by ovipositing females. Nadchatram (1970) noted that larval chiggers (Parasitengona: Trombiculidae) reared from adults fed on a laboratory diet were often paler than larvae of the same species collected in the field. Red female copepods transfer carotenoids to their eggs and thence to their offspring, which are thereby also protected from damaging light (Chalker-Scott 1995). Why might eggs or larval mites need more photoprotection than adults? With a few exceptions (Smith et al. 2001) water mites lay their eggs on floating debris (branches, twigs), partially exposed stones, or vegetation at the water surface. Eggs are encased within a clear gel-like substance and are fixed on the substrate until the larvae hatch. Eggs obviously cannot show behavioral avoidance of light. Water mite larvae vary in their potential exposure to light. Larvae of some species seek their insect hosts below the water's surface, whereas others run about on top of the surface film ('terrestrial larvae') (Smith et al. 2001).

With some exceptions, the water mites that are most consistently red (Hydryphantoidea, Eylaoidea and Hydrachnoidea) have larvae that seek their hosts terrestrially and/or remain on the hosts for a long period of time. Eggs and larvae of these groups are usually also red (H.P., pers. obs.). Larvae of temporary-pool species may remain on their hosts throughout the dry phase of their habitat (weeks to months) (Smith et al. 2001). During aerial searching and much of the time they spend on their hosts, these thin-skinned larvae are exposed to terrestrial light conditions. It is interesting to note that among terrestrial Trombiculidae (Parasitengona: Trombidiina), orange and red larvae run about on open ground, whereas larvae that inhabit nest holes are white or yellow (Nadchatram 1970). Likewise, larval erythraeids (Parasitengona: Erythraeina) found on diurnal opilionids are red, while those on nocturnal Orthoptera are often pale (H.P., pers. obs.).

We therefore suggest that adult mites in ponds may collect carotenoids in order to pass them on to their radiation-bombarded offspring. If so, one might expect that female mites would contain a higher concentration of carotenoids than males. This has been observed in some copepod species (Hairston 1979b) and one species of water mite. Meyer and Kabbe (1991) found that female *Limnochares aquatica* (L.) had higher concentrations of the lutein and canthaxanthin than did conspecific males. However, they also found that males of *Hydrodroma despiciens* (Müller) had much higher concentrations of all carotenoids than did females. Although this appears to be evidence against our hypothesis, it may be that males pass on carotenoids to females via their spermatophores in much the way that male cantharid beetles transfer cantharidin to females in their ejaculate (Chen 1984). The sperm packets produced by male *Hydrodroma* are orange-tinted (H.P., pers. obs.), but there have been no assays for the presence of carotenoids in spermatophores of this or any other taxon of water mite.

Origin of distastefulness

Our laboratory studies showed that powder from terrestrial velvet mites was distasteful to fish, in fact, even more distasteful than the powder from water mites. The apparent aposematism displayed by these red, distasteful terrestrial mites clearly did not evolve in response to fish predation. What is it that makes these mites unpalatable? Both Elton (1923) and Kerfoot (1982) hypothesized that exudates from skin glands were responsible for unpalatability of water mites. Kerfoot (1982) observed that flour balls containing dried secretions from the skin glands of a *Limnesia* species were rejected by guppies. At the moment it is not clear whether a glandular product is also responsible for the apparently unpleasant flavor of *D. magnificum*. Mitchell (1964) does not mention dermal glands in his detailed description of the anatomy of a terrestrial parasitengone. In their review of mite anatomy, Alberti and Coons (1999, p. 727) state that dermal glands are essentially absent in all prostigmatans except for water mites and members of the family Labidostommatidae. Thus it is possible that mechanisms of distastefulness evolved independently in terrestrial and aquatic parasitengones. However, it is also possible that water mite glands simply concentrate chemicals that are also present in the integuments of their terrestrial relatives, and hence distastefulness might also be an ancestral character (see Future Research, below).

In summary, we observed that red individuals dominated the water mite assemblages of temporary (fishless) ponds. These mites also tended to be those that produce terrestrial larvae or larvae that spend weeks to months attached to their hosts. The shallowness of the temporary pond habitat and the habits of the larvae may result in greater exposure to potentially damaging solar radiation. We hypothesize that carotenoid pigments in both terrestrial and aquatic parasitengones serve as photoprotectants, and that the mites' bright colors did not originally evolve as aposematic cues for predators. Relaxed selection for redness has occurred when adults and/or larvae are less exposed to sunlight, either through occupying more protected habitats (e.g. deeper water for aquatic mites, burrows for terrestrial mites), parasitizing more nocturnal hosts, or parasitizing hosts for a shorter period of time. Distastefulness may have evolved after the origin of redness, in order to protect conspicuous mites from attack by visual predators. Most water mites survived repeated attacks by fish (Figure 5) and hence there is no need to use kin selection and/or aggregation to explain the origin of aposematism in this group. The connection between redness and degree of distastefulness observed in water mites by Kerfoot (1982) may also hold for terrestrial parasitengones; however, we did not compare the palatability of red and non-red velvet mites and so can only say that the red species *D. magnificum* is distasteful to fish.

Future research

We present an alternative hypothesis for the evolution of redness and distastefulness in water mites, but do not claim that it is the sole hypothesis. Further study is needed in order to test our suggestion that redness in parasitengone mites is an evolutionary response to damage from solar radiation and that evolution of redness preceded that of distastefulness. First, a better phylogeny of the Parasitengona is required in order to determine the closest terrestrial relatives of water mites. Molecular phylogenies to date have not satisfactorily resolved relationships within this cohort (Otto and Wilson 2001; Söller et al. 2001). If the Hydrachnida represent the ancestral Parasitengona, then it is possible that skin glands have been lost in terrestrial parasitengones but distastefulness retained. If water mites are derived from terrestrial parasitengones, it is also possible that there was only one evolution of distastefulness, and that the original function of dermal glands in water mites was to produce hydrophilic exudates to allow mites to remain under water. The skin glands could have later been co-opted in certain taxa to produce noxious as well as hydrophilic exudates. It would be interesting to expand Kerfoot's (1982) study of distastefulness of glandular secretions across a wider range of water mite taxa, and to correlate degree of distastefulness with variation in structure of the dermal glands (as described in Schmidt 1935).

Second, the prediction that selection for photoprotection acts most strongly on egg and larval and egg life stages should be tested. This could be achieved by subjecting eggs and larvae of congeneric red and non-red species to UV radiation (e.g. as in Ringelberg et al. 1984). Susceptibility to UV damage could be measured both as mortality and as teratogenic effects. Third, behavioral assays of the distastefulness of red terrestrial mites (both Parasitengona and Anystae) should be performed with terrestrial vertebrates as potential predators. It would also be helpful to determine whether visual predators found in fishless water bodies (e.g. salamanders) find red water mites distasteful. If so, then it is possible that pressure from these predators maintain the association between redness and distastefulness in the absence of fish. Fourth, a field test of the assumption that carotenoids are strongly correlated with light exposure experienced by eggs could involve censusing ponds with different light transmissibilities. Water mites inhabiting clearer water bodies or those at higher altitudes should have higher levels of carotenoids than those in turbid, tannin-stained, well-shaded or low-altitude waters where the risk of photodynamic damage is not as great. However, some authorities have observed the opposite pattern, with mites in tannin-stained waters being orange while the same species in clear waters are green or blue (pers. comm., Bruce Smith, Ithaca College). Finally, because the water-body characteristics 'fishless' and 'temporary' were confounded in our study, it would be useful to examine the colors of water mites found in permanent fishless waterbodies.

Acknowledgements

We thank Algonquin Provincial Park, Frank Phelan and the staff of the Queen's University Biological Station for logistic support during this project. Bruce Smith (Ithaca College, New York) and Andy Bohonak (now at San Diego State University) helped with water mite collection and identification. Matt Colloff (CSIRO, Canberra) kindly gave us a photocopy of Dalyell (1851). Bruce Smith and an anonymous referee provided many helpful comments on an earlier draft. Thanks to Justin Schmidt (Carl Hayden Bee Research Center, Tucson, Arizona) for sharing his mite-tasting experiences with us. Research was funded by an NSERC Women's Faculty Award to HP. Work with fish was undertaken with the approval of Queen's University Animal Care Committee.

References

Alberti G. and Coons L.B. 1999. Acari: mites. In: Harrison F.W. and Foelix R.F. (eds), Microscopic Anatomy of Invertebrates, Vol. 8C. Wiley–Liss, New York.

Ang H.P. and Newman L.J. 1998. Warning colouration in pseudocerotid flatworms (Platyhelminthes, Polycladida). A preliminary study. Hydrobiologia 383: 29–33.

Chalker-Scott L. 1995. Survival and sex ratios of the intertidal copepod, *Tigriopus californicus*, following ultra-violet-B (290–320 nm) radiation exposure. Mar. Biol. 123: 799–804.

Chen P.S. 1984. The functional morphology and biochemistry of insect male accessory glands and their secretions. Annu. Rev. Entomol. 29: 233–255.

Czeczuga B. and Czerpak R. 1968a. Pigments occurring in *Hydrachna geographica* and *Piona nodata* (Hydracarina, Arachnoidea). Specialia 24: 218–219.

Czeczuga B. and Czerpak R. 1968b. The presence of carotenoids in *Eylais hamata* (Koenike 1897) (Hydracarina, Arachnoidea). Comp. Biochem. Physiol. 24: 37–46.

Czeczuga B. and Czerpak R. 1968c. Carotenoids in *Hydryphantes dispar* (Schaub, 1988) (Hydracarina, Arachnoidea). Comp. Biochem. Physiol. 25: 547–552.

Elton C.S. 1923. On the colouration of water mites. Proc. Zool. Soc. 82: 1231–1239.

Dalyell J.G. 1851. The Powers of the Creator Displayed in the Creation; or, Observations on Life Amidst the Various Forms of the Humbler Tribes of Animated Nature: with Practical Comments and Illustrations, Vol. 1. John van Voorst, London, UK.

Douglas R.H. and Djamgoz M.B.A. 1990. The Visual System of Fish. Chapman & Hall, New York, USA.

Eckert R., Randall D. and Augustine G. 1988. Animal Physiology: Mechanisms and Adaptations. 3rd ed. W.H. Freeman and Company, New York.

Fox E., Shotton K. and Ulrich C. 1995. SigmaStat Statistical Software User's Manual. Jandel Corporation, San Rafael, CA.

Garga N.G. 1996. Aposematism in water mites (Acari: Hydracarina): a predator defense mechanism, a phylogenetic hold-over, and protection from damaging light. M.Sc. thesis, Queen's University, Kingston, Ontario.

Goodwin T.W. 1984. The Biochemistry of the Carotenoids: 1984, Vol. II. Animals. Chapman & Hall, London.

Green J. 1964. Pigments of the hydracarine *Eylais extendens* (Acari: Hydrachnellae). Comp. Biochem. Physiol. 13: 469–472.

Guilford T. and Cuthill I. 1991. The evolution of aposematism in marine gastropods. Evolution 45: 449–451.

Guilford T. and Dawkins M. 1993. Are warning colours handicaps? Evolution 47: 400–416.

Hairston N.G. Jr. 1979a. The effect of temperature on carotenoid photoprotection in the copepod *Diaptomus nevadensis*. Comp. Biochem. Physiol. 62: 445–448.

Hairston N.G. Jr. 1979b. The relationship between pigmentation and reproduction in two species of *Diaptomus* (Copepoda). Limnol. Oceanogr. 24: 38–44.

Hairston N.G. Jr. 1980. The vertical distribution of diaptomid copepods in relation to body pigmentation. In: Kerfoot W.C. (ed.), Evolution and Ecology of Zooplankton Communities. The University Press of New England, Hanover, pp. 98–110.

Hairston N.G. Jr. 1981. The interaction of salinity, predators, light and copepod color. Hydrobiologia 81: 151–158.

Harvey M.S. 1998. The Australian Water Mites: A Guide to Families and Genera. Monographs on Invertebrate Taxonomy 4. CSIRO Publishing, Collingwood.

Huhta E., Rytkönen S. and Solonen T. 2003. Plumage brightness of prey increases predation risk: an among-species comparison. Ecology 84: 1793–1799.

Joron M. 2003. Aposematic coloration. In: Resh V.H. and Cardé R.T. (eds), Encyclopedia of Insects. Academic Press, Amsterdam, pp. 39–45.

Kerfoot W.C. 1982. A question of taste: crypsis and warning coloration in freshwater zooplankton communities. Ecology 63: 538–554.

Krantz G.W. 1978. A Manual of Acarology, 2nd ed. Oregon State University Book Stores, Inc. Corvallis, USA.

Leimar O., Enquist M. and Sillén-Tullberg B. 1986. Evolutionary stability of aposematic colouration and prey uprofitability: a theoretical analysis. Am. Nat. 128: 469–490.

Lindström L. 1999. Experimental approaches to studying the initial evolution of conspicuous aposematic signalling. Evol. Ecol. 13: 605–618.

Luecke C. and O'Brien W.J. 1981. Phototoxicity and fish predation: selective factors in color morphs of *Heterocope*. Limnol. Oceanogr. 26: 454–460.

Manunta C. 1939. Estraazione e cristallizzazione del pigmento che colora in rosso la pelle di certi acari del genere Trombidium. Helv. Chim. Acta. 22: 1154–1155.

Merilaita S. and Kaitala V. 2002. Community structure and the evolution of aposematic coloration. Ecol. Lett. 5: 495–501.

Metcalf R.L. and Newell I.M. 1962. Investigation of the biochromes of mites. Ann. Entomol. Soc. Am. 55: 350–353.

Meyer E. and Kabbe K. 1991. Pigmentation in water mites of the genera *Limnochares* Latr. and *Hydrodroma* Koch (Hydrachnidia). In: Schuster R. and Murphy P.W. (eds), The Acari: Reproduction, Development and Life-History Strategies. Chapman & Hall, New York, USA, pp. 379–391.

Mitchell R. 1964. The anatomy of an adult chigger mite *Blankaartia acuscutellaris* (Walch). J. Morphol 114: 373–391.

Nadchatram M. 1970. Correlation of habitat, environment and color of chiggers, and their potential significance in the epidemiology of scrub typhus in Malaya (Prostigmata: Trombiculidae). J. Med. Entomol. 7: 131–144.

N.H. Analytical Software. 1989. Statistix 3.1. Roseville, Minnesota.

Otto J.C. and Wilson K.J. 2001. Assessment of the usefulness of ribosomal 18S and mitochondrial COI sequences in prostigmata phylogeny. In: Halliday R.B., Walter D.E., Proctor H.C., Norton R.A. and Colloff M.J. (eds), Acarology: Proceedings of the 10th International Congress. CSIRO Publishing, Melbourne, pp. 100–109.

Pearse V., Pearse J., Buchsbaum M. and Buchsbaum R. 1987. Living Invertebrates. Blackwell Scientific Publications, Pacific Grove.

Pennak R.W. 1976. Fresh-water Invertebrates of the United States, 2nd ed. John Wiley & Sons, New York.

Ringelberg J., Keyser A.L. and Flik B.J.G. 1984. The mortality effect of ultraviolet radiation in a translucent and in a red morph of *Acanthodiaptomus denticornis* (Crustacea, Copepoda) and its possible ecological relevance. Hydrobiologia 112: 217–222.

Schmidt U. 1935. Beiträge zur anatomie und histologie der hydracarinen, besonders von *Diplodontus despiciens*. O.F. Müller. Zeit. Morph. Ökol. Tiere 30: 99–176.

Scott W.B. and Crossman E.J. 1973. Freshwater Fishes of Canada. Fisheries and Research Board of Canada, Bulletin 184, Information Canada, Ottawa.

Sillén-Tullberg B. 1988. Evolution of gregariousness in aposematic butterfly larvae: a phylogenetic analysis. Evolution 42: 293–305.

Smith B.P. 1990. Hydrachnidia. In: Peckarsky B.L., Fraissinet P.R., Penton M.A. and Conklin D.J. (eds), Freshwater Macroinvertebrates of Northeastern North America. Cornell University Press, Ithaca, pp. 290–334.

Smith I.M., Cook D.R. and Smith B.P. 2001. Water mites (Hydrachnida) and other arachnids. In: Thorp J.H. and Covich A.P. (eds), Ecology and Classification of North American Freshwater Invertebrates, 2nd ed. Academic Press, San Diego, pp. 551–659.

Söller R., Wohltmann A., Witte H. and Blohm D. 2001. Phylogenetic relationships within terrestrial mites (Acari: Prostigmata, Parasitengona) inferred from comparative DNA sequence analysis of the mitochondrial cytochrome oxidase subunit I gene. Mol. Phylogenet. Evol. 18: 47–53.

Summers K. and Clough M.E. 2001. The evolution of coloration and toxicity in the poison frog family (Dendrobatidae). Proc. Natl. Acad. Sci. 98: 6227–6232.

Tullberg B.S. and Hunter A.F. 1996. Evolution of larval gregariousness in relation to repellent defences and warning coloration in tree-feeding Macrolepidoptera: a phylogenetic analysis based on independent contrasts. Biol. J. Linn. Soc. 57: 253–276.

Tullberg B.S., Leimar O. and Stille G. 2000. Did aggregation favour the initial evolution of warning coloration? A novel world revisited Anim. Behav. 59: 281–287.

Turner J.R.G. 1971. Studies of Müllerian mimicry and its evolution in burnet moths and heliconid butterflies. In: Creed R. (ed.), Ecological Genetics and Evolution. Blackwell Publishing, Oxford, pp. 224–260.

Welbourn W.C. 1991. Phylogenetic studies of the terrestrial Parasitengona. In: Dusbabek F. and Bukva V. (eds), Modern Acarology, Vol. 2. SPB Academic Publishing bv, The Hague, The Netherlands, pp. 163–170.

Witte H. 1991. The phylogenetic relationships within the Parasitengonae. In: Dusbabek F. and Bukva V. (eds), Modern Acarology, Vol. 2. SPB Academic Publishing bv, The Hague, The Netherlands, pp. 171–182.

Zar J.H. 1996. Biostatistical Analysis, 3rd ed. Prentice Hall International, Inc., Upper Saddle River, NJ.

Zhang Z.-Q. 1998. Biology and ecology of trombidiid mites (Acari: Trombidioidea). Exp. Appl. Acarol. 22: 139–155.

Experimental and Applied Acarology **34:** 149–169, 2004.

Of spates and species: responses by interstitial water mites to simulated spates in a subtropical Australian river

ANDREW BOULTON[1,*], MARK HARVEY[2] and HEATHER PROCTOR[3]

[1]*Ecosystem Management, University of New England, Armidale, New South Wales 2350, Australia;* [2]*Department of Terrestrial Invertebrates, Western Australian Museum, Francis St, Perth, Western Australia 6000, Australia;* [3]*Department of Biological Sciences, University of Alberta, Edmonton, Alberta, Canada T6G 2E9; *Author for correspondence (e-mail: aboulton@metz.une.edu.au; phone: +61-2-6773-3760; fax: +61-2-6773-2769)*

Key words: Australia, Flow refuges, Hyporheic refuge hypothesis, Interstitial water mites, Spates, Subtropical river

Abstract. The 'hyporheic refuge hypothesis' predicts that the hyporheic zone, the saturated sediments below and alongside rivers and streams, is a refuge from the scouring effects of spates for many aquatic invertebrates including water mites. We tested this hypothesis in two lateral gravel bars and two riffles in a subtropical Australian river by collecting water mites from the hyporheic zone at two depths (10 and 50 cm) at two'pre-flood' sampling times before experimentally diverting water through the sites for 14 h to simulate a spate. Taxon richness of mites washigh (46 taxa) and dominated by the Prostigmata, with nearly half the species being new to science. Oribatids were also common at the four sites. Samples were collected twice during each 'spate', and again soon after flow was returned to normal. The experimental spate induced changes in the strength and even direction of subsurface–surface water exchange; however, these changes seldom persisted after the experiment, nor after a subsequent natural spate. The hyporheic refuge hypothesis was not supported by our water mite data. Neither during nor shortly after the experimental spates did we find more epigean (surface-dwelling) water mites in downwelling zones where surface streamwater enters the hyporheic zone, demonstrating that these mites were not using the hyporheic zone as a refuge at these locations. There was also no evidence for a 'wash out' effect, because hyporheic mitedensities did not significantly decline late in the spate. Our data indicate that floods of the low magnitude simulated in this study apparently do not pose a lasting disturbance for hypogean water mites. The fact that the same response was found at four sites indicates that the hyporheic refuge hypothesis may not always be an appropriate explanation for rapid post-flood recolonisation. Possibly, the use of the hyporheic zone as a refuge from floods may be dictated by the strength of the disturbance and substrate composition and stability.

Introduction

The 'hyporheic refuge hypothesis' posits that the hyporheic zone, the saturated sediments below and alongside river channels, provides a safe refuge for many aquatic invertebrates from natural environmental disturbances such as spates and droughts (review in Williams and Hynes 1974). Much of the hyporheic zone resists physical disturbance during spates (Dole-Oliver et al. 1997), and for aquatic invertebrates small enough to be able to penetrate the interstices among sediment particles, it is predicted that there is a refuge from scour and high water flows. At first glance, this hypothesis seems plausible and it has been echoed by many authors (e.g., Williams 1984; Triska et al. 1989; Marmonier and Creuzé des Châtelliers 1991; Richards and Bacon 1994; Brunke and Gonser 1997; Boulton 2000). Even lotic water mites have been suggested to find refuge in the interstitial zone during unfavourable surface conditions such as spates (Di Sabatino et al. 2000). However, the flood refuge hypothesis remains untested for most taxa despite its widespread citation and implications for river ecology and management.

Several studies imply that the hyporheic zone may not be such a universal refuge. Working in a sand-bed stream in Virginia, Palmer et al. (1992) reported different degrees of use of the hyporheic zone by stream meiofauna as a refuge from spates. This difference in response was partly attributed to the limited ability of many stream invertebrates to enter the small spaces among the sand grains and the restricted physical heterogeneity of the stream bed. In France, Dole-Oliver et al. (1997) also demonstrated the patchy usage of the hyporheic zone as a refuge from fifteen natural spates varying in magnitude from 50 to 1640 m^3/s that flowed down a by-passed section of the Rhône River, where the sediments are large and the hyporheic zone is highly permeable. Because of the unpredictable occurrence of spates, their field study relied on sampling the hyporheic zone after the events had occurred, and interpretation was hampered by lack of pre-spate data (Dole-Olivier et al. 1997).

Boulton and Harvey (2003) sought to test the hyporheic refuge hypothesis in a field experiment by sampling the water mite fauna of the hyporheic zone of an Australian subtropical river before, during, and after a simulated spate. Their results from only a single site failed to support the flood refuge hypothesis, and they concluded that for the hyporheic water mites in this subtropical river, floods of the low magnitude simulated in their study apparently do not pose a lasting disturbance. Epigean water mites also did not seem to migrate actively into the hyporheic zone to avoid the increased surface flows. However, the study was limited by being restricted to only a single lateral bar (Boulton and Harvey 2003) and it was felt that the hypothesis would be better tested by enlarging the data set to include an additional bar as well as the hyporheic zones below two riffles to see if the responses were consistent across sites.

Water mites were chosen as a target group to test the hyporheic refuge hypothesis for several reasons. Firstly, water mites have epigean (surface) and interstitial members so that changes in densities of these 'functional groups' (*sensu* Claret et al. 1999) could be compared in response to the spate. Secondly, water mites are especially diverse in the hyporheic zone of the study river (Boulton and Foster 1998; Boulton and Harvey 2003) and their taxonomy is reasonably well-known (e.g. Cook 1986; Harvey 1990, 1998a, b; Smit 1998; Harvey 2003). Thirdly, the behavioural ecology of lotic water mites, especially interstitial ones, is virtually unknown (Di Sabatino et al. 2000) and most studies of the hyporheic zone seldom differentiate water mite species. Although water mite assemblages have been used to define the typology of springs and other groundwaters (Schwoerbel 1959; Smith 1991; Gerecke and Di Sabatino 1996), most studies of the hyporheic zone have not identified water mites to species or even genus level, preventing detection of species-specific responses.

We predicted that if the hyporheic refuge hypothesis is correct, during and shortly after a spate, more epigean water mites would be found in the hyporheic zone, especially in downwelling zones where surface streamwater enters the hyporheic zone. An alternative possibility could be that, given the high permeability of the sediments in the river we studied, there may be a 'wash out' effect (Marmonier and Creuzé des Châtelliers 1991) that might nullify the hyporheic refuge for many invertebrates. Therefore, if this 'wash out' effect was taking place, we hypothesised that total hyporheic water mite densities would decline late in the spate, even in upwelling zones where hyporheic water enters the surface stream. Soon after our experiment, the fortuitous occurrence of a natural flood provided limited comparative data to test our hypotheses at the two lateral bars.

Methods and materials

Study area

The experiments were conducted in two riffles (R1, R2) and along their adjacent lateral bars (B1, B2) near Tallowood Point (30°21′S, 152°54′E) on the Never Never River, a subtropical gravel-bed river in northern New South Wales, Australia. Upstream of the site, the catchment lies in National Park (970 m a.s.l.) and is heavily vegetated with rainforest and, in some areas, plantation eucalypts (*Eucalyptus saligna*). Water quality is excellent and conductivity is low (ca. 30–50 μS/cm). Further details of the site and its catchment characteristics are in Boulton and Foster (1998). Mean annual rainfall is 1497 mm, with most falling in austral summer (January–March) (Boulton and Harvey 2003).

Methodology

Experiments were conducted in December 2000 after a period of some 5 months without significant flooding. At that time, discharge was 1200 l/s, determined as the product of cross-sectional area and multiple measures of velocity (Gordon et al. 1992). Artificial spates were generated by building temporary barriers of thick plastic sheeting supported on a fence of chicken wire in the stream to either funnel the water down the riffle in a narrow channel or to divert water across the lateral gravel bars on one side of the channel. Each dam was capable of raising water height by 30–50 cm, increasing riffle surface flow by at least threefold, and inundating 45–65 m^2 of the lateral bars (full details of barrier construction are in Boulton and Harvey 2003). In the riffles, there was some movement of bed sediments during the artificial spates but sediments of the lateral bars were not disturbed. A natural spate in late December 2000 inundated the lateral bars to approximately the same extent as the experimental treatments. This flood receded after 10–12 h, allowing us to collect post-spate samples from the lateral bars.

At each of the two study riffles, downwelling and upwelling sites were identified at the head and base of the riffle, respectively. Along the two lateral bars, three sites were chosen: a downwelling site where stream water entered the upstream end of the bar, a second site at the end of a 'short' subsurface flowpath, and a third site where a 'long' subsurface flowpath ended. At each site, three pairs of PVC wells (15 mm internal diameter) were inserted to depths of 10 and 50 cm. Five sets of samples were taken during the experiment: two pre-spate (T1 and T2), two during the spate (T3 and T4), and a single sample post-spate (T5) (Table 1). Two days after the natural spate, a sixth set of samples was collected (T6) at both bars but these could not be collected in the riffles.

Six litres of hyporheic water were pumped out of each well using a manual bilge pump (method described in Boulton et al. 1992) and filtered through a 125 μm mesh to collect the invertebrates and particulate organic matter. The degree of upwelling or downwelling of water at each well was measured using a transparent Plexiglass tube (piezometer) inserted into the well to 10 or 50 cm before sliding the PVC well out. This enabled comparison of the vertical difference between the stream water level and the piezometric level of water in the Plexiglass tube, expressed as a vertical head (Boulton 1993; Boulton and Foster 1998). Faunal samples were stained with Rose Bengal and preserved in ethanol for return to the laboratory where they were sorted under 32× magnification. Water mites were preserved in Koenike's Fluid and identified to species level. All specimens are lodged in the Western Australian Museum, Perth. Prior to analysis, taxa were ascribed to 'epigean' and 'hypogean' categories based on their morphology. Hypogean mites generally possess a suite of morphological adaptations to an interstitial existence, including loss of pigmentation, loss of eye capsules and/or eye spots and loss of swimming hairs (Schwoerbel 1967; Di Sabatino et al. 2000).

After testing the data for normality (Shapiro–Wilk test, Analytical Software 2000) and transforming dependent variables where required, differences in hydraulic head and mean densities of epigean and hypogean water mites were compared between depth, site, and over time using three-way Analyses of Variance (ANOVAs). The triplicate wells served as replicates. We tested the hyporheic refuge hypothesis by searching for significant interaction effects of mean densities of epigean and hypogean water mites in each habitat (site and depth) over time. We compared mean rank densities across the time-depth combinations within habitats using Kruskal–Wallis tests followed by Bonferroni comparisons as above. If the hyporheic refuge hypothesis was supported,

Table 1. Water mites recorded from two riffles (R1, R2) and their adjacent lateral bars (B1, B2) in the Never Never River, New South Wales, Australia.

		R1	R2	B1	B2
Epigean taxa					
Oribatida		*	*	*	*
Prostigmata					
Momoniidae	*Momoniella parva* Cook		*		*
Oxidae	*Frontipoda* sp.	*		*	*
	Flabellifrontipoda sp.	*			
	Oxus sp.	*		*	*
Aturidae	*Albia lundbladi* Cook		*		
	Austraturus vietsi Cook		*		
	Austraturus sp. nov.	*	*	*	
Hygrobatidae	*Aspidiobates* sp. nov.			*	
	Aspidiobates unidorsalis Smit	*			
	Kallimobates sp.				*
	Procorticacarus cooki (Imamura)	*	*	*	*
	Procorticacarus sp.		*		*
	Australorivacarus sp.		*		*
	Rhynchaustrobates sp.		*		
Limnesiidae	*Tubophorella amoena* K.O. Viets			*	
Unionicolidae	*Koenikea* sp. nov. (nr *lemba* Cook)			*	
Torrenticolidae	*Monatractides* sp. nov.	*	*	*	
Trombididioidea					*
Mesostigmata				*	*
Astigmata					*
Hyporheic taxa					
Prostigmata					
Aturidae	*Twarntaturus* sp. nov. (nr *australicus* Cook)	*	*	*	*
	Thryptaturus boultoni Harvey				*
	Notoaturinae gen. nov.		*	*	*
	Notoaturinae gen. nov. A (*sensu* Harvey)			*	
	Cabellaturus sp. nov.	*	*	*	*
	Spinaturus ctenophorus Cook	*	*		
	Barwontius sp. nov.	*		*	*
	Axonopsella sp.	*	*	*	*
Halacaridae	Gen. nov.	*			
	Lobohalacarus sp.			*	*
	Soldanellonyx monardi (Walter)	*	*	*	*
Hydryphantidae	*Wandesia (Pseudowandesia)* sp. nov.		*		*
Omartacaridae	*Omartacarus* sp. nov.	*	*	*	*
Anisitsiellidae	*Anisitsiellides* sp. nov.				*
	Anisitsiellides circularis Cook		*	*	
	Rutacarus sasonus Cook			*	*
Athienemanniidae	*Mellamunda* sp. nov.	*	*	*	*
	Davecookia peramica Harvey			*	
Pezidae	*Peza* sp. nov.	*			*
Momoniidae	*Partidomomonia* sp. 'c'	*	*	*	*
Mideopsidae	*Penemideopsis* sp. nov.	*		*	
	Penemideopsisphreatica Cook	*	*	*	*
	Guineaxonopsis sp. nov.				*
Hygrobatidae	*Australiobates* sp. nov.	*	*	*	*
Frontipodopsidae	*Frontipodopsis* sp. nov.			*	

* = present

Taxa were allocated to 'epigean' or 'hypogean' categories based on their morphology and the literature.

we predicted significantly more epigean mites in downwelling zones soon after the spate. If a 'wash out' effect occurred, overall mite densities should decline during the spate. Depending on recolonisation rates, this decrease was hypothesised to persist after the experimental (T5) or natural (T6) spates.

Results

Hydrological responses to the simulated and real spates

In both riffles, surface water entered the hyporheic zone in the downwelling zones with negative hydraulic heads at the tops of the channels and then upwelled (positive hydraulic head) at the lower end. This pattern persisted during the experiment (Figures 1 and 2). In the downwelling zones of both riffles, the simulated spate significantly increased the negative hydraulic heads ($F_{7,23} = 44.02$, $p < 0.0001$ at R1; $F_{7,23} = 18.29$, $p < 0.0001$ at R2) by two- to threefold (Figures 1 and 2), demonstrating the effectiveness of the experimental treatment in enhancing streamflow into the sediments. However, there

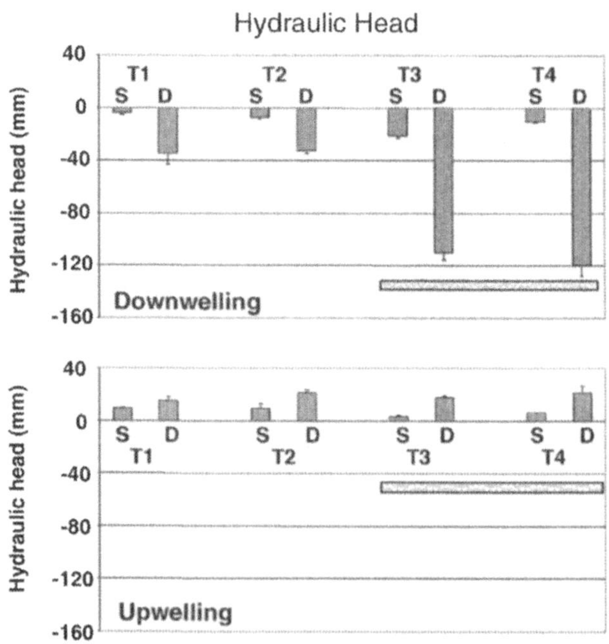

Figure 1. Mean (+1 SE) hydraulic head (n = 3 wells) at 10 cm depth (S = Shallow) and 50 cm depth (D = Deep) in the first riffle R1 before (T1, T2) and during (T3, T4) an experimental spate at the downwelling (top panel) and upwelling (bottom panel) zones. Negative values indicate downwelling water whereas positive values imply upwelling water (positive head). The hatched bar represents the high flow period.

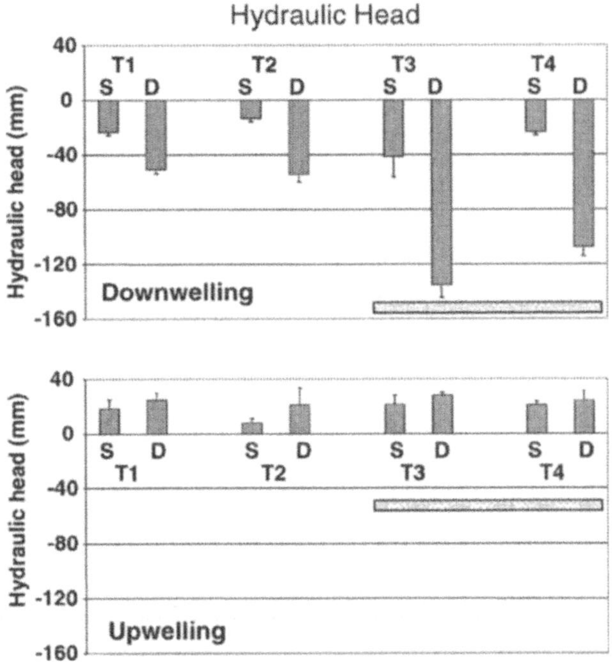

Figure 2. Mean (+1 SE) hydraulic head ($n = 3$ wells) at 10 cm depth (S = Shallow) and 50 cm depth (D = Deep) in the second riffle R2 before (T1, T2) and during (T3, T4) an experimental spate at the downwelling (top panel) and upwelling (bottom panel) zones. Negative values indicate downwelling water whereas positive values imply upwelling water (positive head). The hatched bar represents the high flow period.

was no significant increase in the strength of upwelling in either riffle ($p > 0.05$) although the difference in hydraulic head with depth persisted. Unfortunately, there was no opportunity to sample these riffles soon after the experimental spate as a real spate occurred, restricting post-spate sampling to the two lateral bars.

Hydrological responses in the lateral bars to the experimental spates were more complex and differed between the two sites. At B1, the experimental spate induced a significant increase ($F_{11,35} = 15.45$, $p < 0.0001$) in the strength of the downwelling of the deeper wells similar to the effect noted in both riffles, but the hydraulic heads returned to normal soon after and also recovered quickly after the natural spate (Figure 3). However, at B2, there seemed to be no substantial hydrological response in the downwelling zone during the artificial spate yet soon after, there was significantly stronger downwelling, especially in the −10 cm wells ($F_{11,35} = 19.95$, $p < 0.0001$, Figure 4). After the natural spate, the patterns of downwelling heads were not significantly different from those before or during the experimental spate ($p > 0.05$, Figure 4).

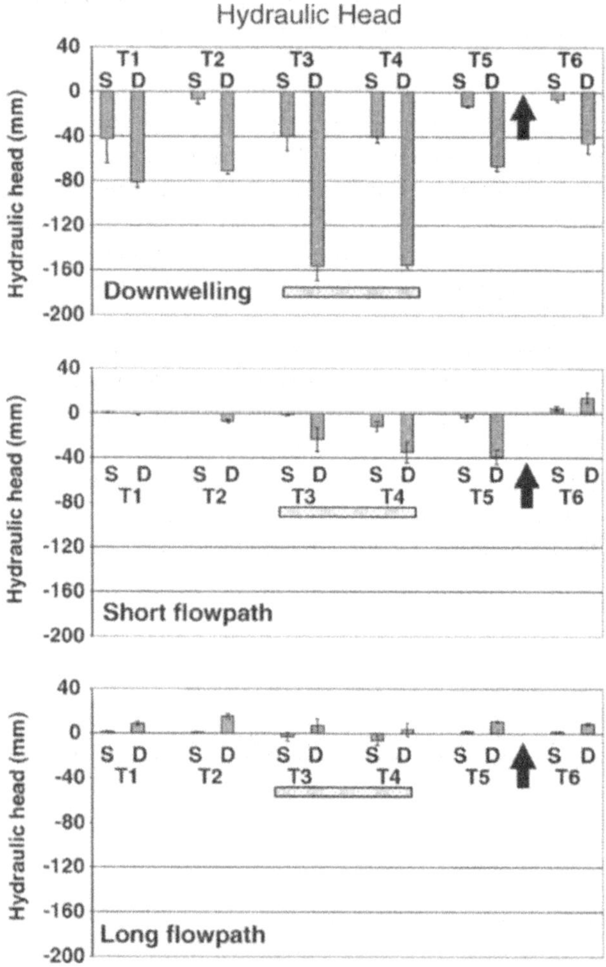

Figure 3. Mean (+1 SE) hydraulic head (n = 3 wells) at 10 cm depth (S = Shallow) and 50 cm depth (D = Deep) in the first bar B1 before (T1, T2), during (T3, T4) and after the experimental spate (T5) and after a natural spate (T6) in the downwelling zone at the head of the bar (top panel), terminus of the short flow path (middle panel) and terminus of the long flow path (bottom panel). Negative values refer to downwelling water whereas positive values imply upwelling water (positive head). The hatched bar represents the high flow period, the arrow indicates the natural spate.

Vertical hydrological exchange was limited at the end of the short flowpath at B1 before the experimental spate but became significantly downwelling in the deeper wells ($F_{11,35}$ = 7.88, p < 0.0001), an effect that persisted after the experiment (Figure 3). When we sampled after the natural flood (T6), this zone was now an upwelling one, implying that transiently stored floodwater was leaving the shallow alluvial aquifer at this point. At the end of the long flowpath at the same site, the hydraulic heads of shallow and deep wells were

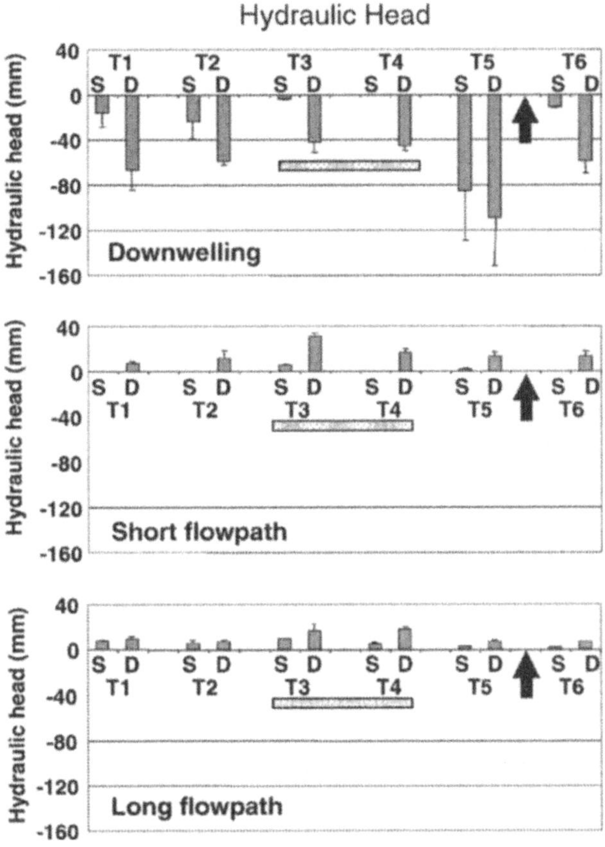

Figure 4. Mean (+1 SE) hydraulic head (n = 3 wells) at 10 cm depth (S = Shallow) and 50 cm depth (D = Deep) in the second bar B2 before (T1, T2), during (T3, T4) and after the experimental spate (T5) and after a natural spate (T6) in the downwelling zone at the head of the bar (top panel), terminus of the short flow path (middle panel) and terminus of the long flow path (bottom panel). Negative values refer to downwelling water whereas positive values imply upwelling water (positive head). The hatched bar represents the high flow period, the arrow indicates the natural spate.

typically positive although this declined and even, in the shallow wells, became negative during the experimental spate (Figure 3). There was no significant change in response to the real spate ($p > 0.05$). Conversely, at B2, the ends of both flowpaths were upwelling zones where the strength of the hydraulic head increased significantly in the deeper wells during the experimental flood ($F_{11,35} = 17.99$, $p < 0.0001$) but this effect did not persist, even after the natural spate (Figure 4).

Water mite assemblages in two riffles and lateral bars of a subtropical river

In total, 988 water mites representing 46 taxa were collected during this experiment (Table 1). About one-fifth of these specimens were not mature enough to identify confidently to species or genus level and were not included in subsequent analyses. Prostigmata dominated taxon richness with 43 species, of which at least half appear to be new species or have been described only recently (e.g., *Davecookiaperamica* Harvey 2003). The most common prostigmatans were the epigean hygrobatid *Procorticacarus cooki* (Imamura) and new species of hypogean aturids in the genera *Twarntaturus*, *Cabellaturus*, and *Axonopsella*. Also relatively common were the hypogean omartacarid *Omartacarus* sp. nov. and the mideopsid *Penemideopsis phreatica* Cook. Although oribatids were not identified beyond suborder in this study, at least three different species were apparent in the genera *Malaconothrus* and *Trimalaconothrus* (Malaconothridae) and *Mucronothrus* (Trhypochthoniidae). Oribatids were common at all sites and habitats before, during and after the spates whereas Astigmata and Mesostigmata were rare and potentially accidental 'fall ins'.

Overall, representation of most species was extremely patchy across the sites and 19 of the 46 taxa occurred at only one of the four habitats (Table 1), often as only 1–2 individuals. Abundances were also highly variable and for that reason, little information could be gleaned from statistical analyses of even the more common taxa. Therefore, analysis was confined to the two groups classified according to their general morphology as epigean or hypogean. Approximately half the taxa were epigean.

Responses of epigean water mites to artificial and real floods

The hypothesis that epigean water mite densities would rise in the downwelling hyporheic zone as they sought refuge from the experimental spate was not supported in the two riffles. There was no significant change in densities of epigean mites in the downwelling zone at R1 ($H_{7,23} = 5.2402$, $p = 0.6307$) or at R2 ($H_{7,23} = 7.7719$, $p = 0.3531$), nor was there a significant difference in densities between the deep and shallow wells at each riffle ($p > 0.05$, Figures 5 and 6). Mean densities ranged between 2 and 6 epigean water mites per sample, except in the shallow wells at R2 (mean of 13 mites, Figure 6) where high numbers of epigean oribatids occurred in a single sample. Similarly, there were no significant differences in epigean mite densities in the upwelling zones before and during the experimental spate in both riffles ($p > 0.05$, Figures 5 and 6), nor were there any differences in densities between well depths in this zone at R1 or R2.

At the B1 lateral bar, there were significant differences in densities of epigean water mites among habitats ($F_{2,72} = 348.09$, $p < 0.0001$), depths ($F_{1,72} = 139.39$, $p < 0.0001$) and over time ($F_{5,72} = 30.30$, $p < 0.0001$). Epigean mites were more common in shallow than deep wells in the downwelling zone before and during the experimental spate but these differences became non-significant after the experimental spate and also after the natural spate ($p > 0.05$, Figure 7). There was no evidence of increased densities during or after the experimental spate, disproving the hyporheic refuge hypothesis. At

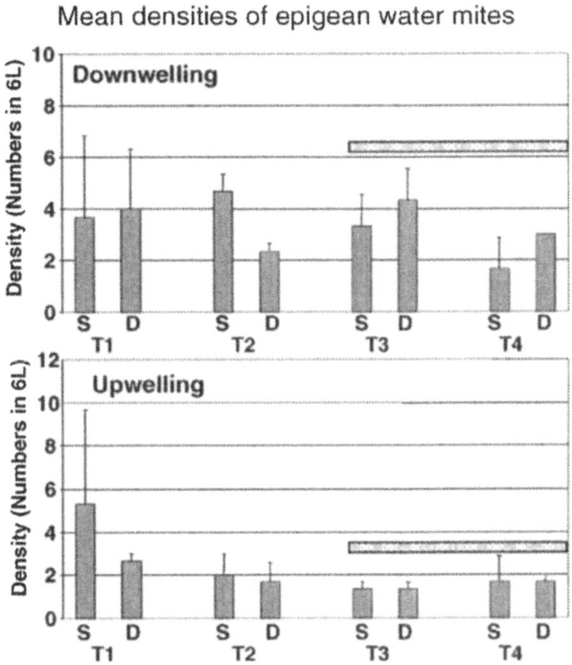

Figure 5. Mean (+1 SE) epigean mite densities (n = 3 wells) at 10 cm depth (S = Shallow) and 50 cm depth (D = Deep) in the first riffle R1 before (T1, T2) and during (T3, T4) an experimental spate at the downwelling (top panel) and upwelling (bottom panel) zones. The hatched bar represents the high flow period.

the end of the short flowpath, densities of epigean water mites in the deep wells were significantly lower during the experimental spate and after the natural spate ($F_{11,24} = 2.64$, $p = 0.0228$) whereas at the end of the long flowpath at B1, mean epigean mite densities were typically low (Figure 7) and did not vary significantly over time.

In contrast, at B2 there were no significant differences in epigean mite densities among habitats or depths ($p > 0.05$, Figure 8). However, after 14 h of experimental spate (T4) and after the natural spate (T6), epigean mite densities were significantly lower ($F_{11,24} = 2.81$, $p = 0.0241$) across all habitats and depths, apparently indicating a 'wash-out' effect. On closer inspection, this difference was found not to hold true within habitats, and there were no significant differences in epigean mite densities over time at either depth in each habitat ($p > 0.05$, Figure 8). At both bars, mean densities of epigean water mites were typically lowest at the ends of the long flowpaths (Figures 7 and 8).

Responses of hypogean water mites to artificial and real floods

Densities of hypogean water mites did not vary significantly over time in downwelling or upwelling zones at either riffle site ($p > 0.05$, Figures 9 and 10)

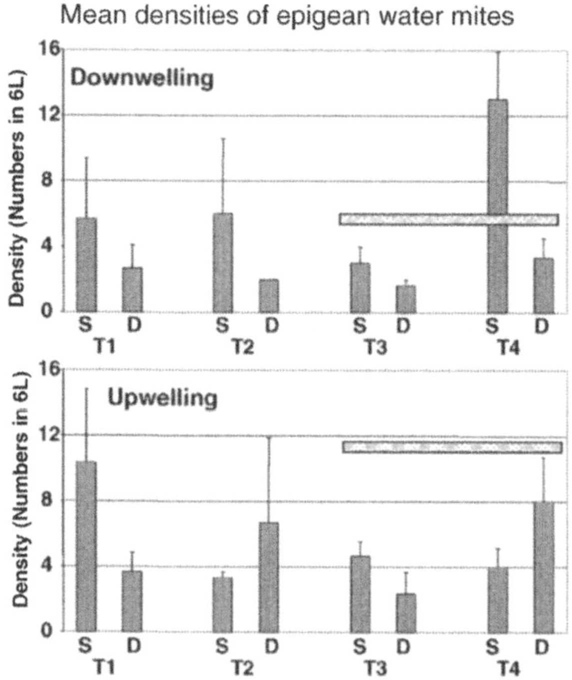

Figure 6. Mean (+1 SE) epigean mite densities ($n = 3$ wells) at 10 cm depth (S = Shallow) and 50 cm depth (D = Deep) in the second riffle R2 before (T1, T2) and during (T3, T4) an experimental spate at the downwelling (top panel) and upwelling (bottom panel) zones. The hatched bar represents the high flow period.

although there were significantly more hypogean water mites in the upwelling than in the downwelling zone at R2 ($F_{1,32} = 4.65$, $p = 0.0386$), especially soon after the experimental spate commenced (T3, Figure 10). Although this might imply some immediate movement of hypogean mites in response to changes in hydrology, there was no marked concomitant increase in hydraulic head in the upwelling zone at this site when the experimental spate commenced (Figure 2). There was no evidence for a 'wash out' effect of hypogean water mites at either of these riffles.

In the bar at B1, there was no significant difference in hypogean water mite densities in the downwelling zone or at the end of the long flowpath ($p > 0.05$, Figure 11). However, overall densities of hypogean mites declined at the end of the short flowpath during the experimental spate, and remained low in the deep wells (Figure 11). This pattern corresponded with the transient increase in downwelling hydraulic head during the experimental spate that was especially marked in the deep wells (Figure 3). However, after the natural flood, this zone changed to a weakly upwelling one (Figure 3).

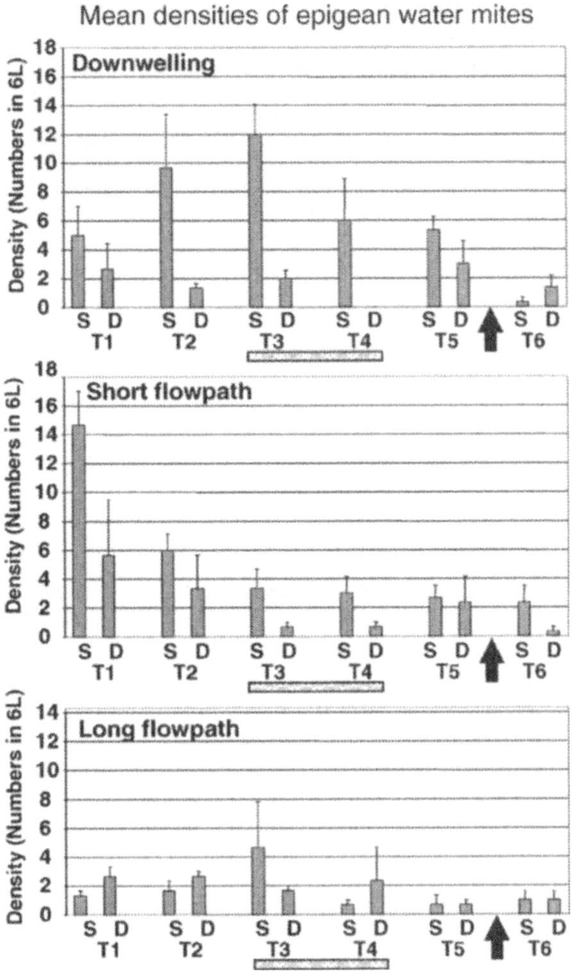

Figure 7. Mean (+1 SE) epigean mite densities ($n = 3$ wells) at 10 cm depth (S = Shallow) and 50 cm depth (D = Deep) in the first bar B1 before (T1, T2), during (T3, T4) and after the experimental spate (T5) and after a natural spate (T6) in the downwelling zone at the head of the bar (top panel), terminus of the short flow path (middle panel) and terminus of the long flow path (bottom panel). The hatched bar represents the high flow period, the arrow indicates the natural spate.

At B2, hypogean water mites were significantly more abundant in the deeper wells ($F_{1,72} = 12.80$, $p = 0.0002$), and this was clearly evident in the downwelling zone and the end of the short flowpath at most times (Figure 12). As this pattern existed before the experimental spates and did not change during the experiment, it can be assumed that there was no 'wash out' effect and that the hypogean water mites showed no obvious response to artificial spate. There was also little response to the natural flood (Figure 12).

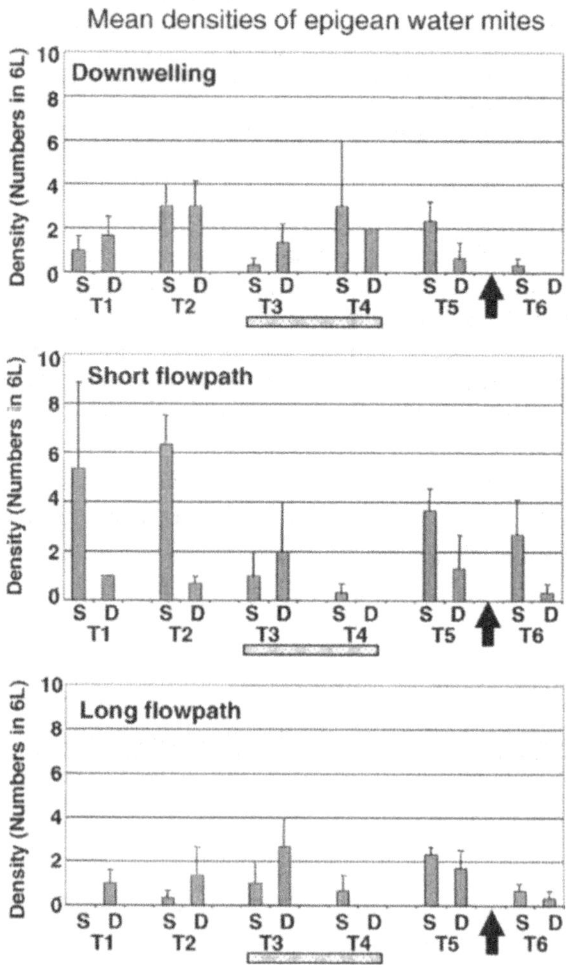

Figure 8. Mean (+1 SE) epigean mite densities (n = 3 wells) at 10 cm depth (S = Shallow) and 50 cm depth (D = Deep) in the second bar B2 before (T1, T2), during (T3, T4) and after the experimental spate (T5) and after a natural spate (T6) in the downwelling zone at the head of the bar (top panel), terminus of the short flow path (middle panel) and terminus of the long flow path (bottom panel). The hatched bar represents the high flow period, the arrow indicates the natural spate.

Discussion

Our experimental and field data from the hyporheic zones of two riffles and two lateral bars in this subtropical river do not support the predictions of the hyporheic refuge hypothesis (Williams and Hynes 1974). At all habitats sampled, there was no increase in the densities of epigean water mites in the downwelling zones after the experimental spate commenced. If epigean mites

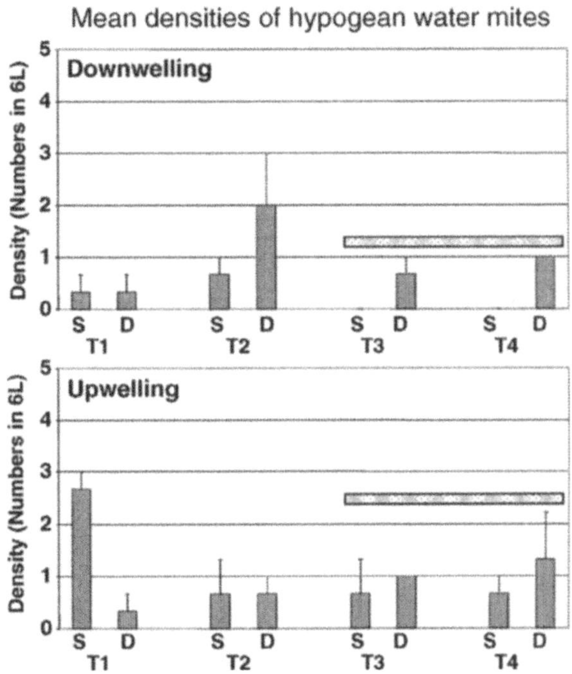

Figure 9. Mean (+1 SE) hypogean mite densities (n = 3 wells) at 10 cm depth (S = Shallow) and 50 cm depth (D = Deep) in the first riffle R1 before (T1, T2) and during (T3, T4) an experimental spate at the downwelling (top panel) and upwelling (bottom panel) zones. The hatched bar represents the high flow period.

were seeking refuge from the experimental spates, they did not appear to be accumulating in the downwelling zones nor was there a trend over time for more epigean mites to be found at the ends of the short or long flow paths. Hypogean mites tended to be more abundant in the deeper wells in the lateral bars as expected but did not seem to show strong depth preferences in the riffle. As for epigean mites, their densities across the four habitats did not change significantly in response to the experimental spates. Thus, the results of our study concur with those by Palmer et al. (1992) and Dole-Olivier et al. (1997) in other streams, and despite expectations in the literature (e.g., Di Sabatino et al. 2000) cast further doubt on the generality of the hyporheic refuge hypothesis for water mites.

Although there was hydraulic evidence of enhanced inflow of streamwater in the downwelling zones in the riffles during the experiments, there was no associated increase in the magnitude of upwelling, implying that the water was perhaps being dissipated laterally. This is important because we were unable to demonstrate any ‘wash out’ effect as observed by Marmonier and Creuzé des Châtelliers (1991) where hyporheic invertebrates were flushed by floods from gravel bars of the Rhone River. The sediments of the Never Never River are

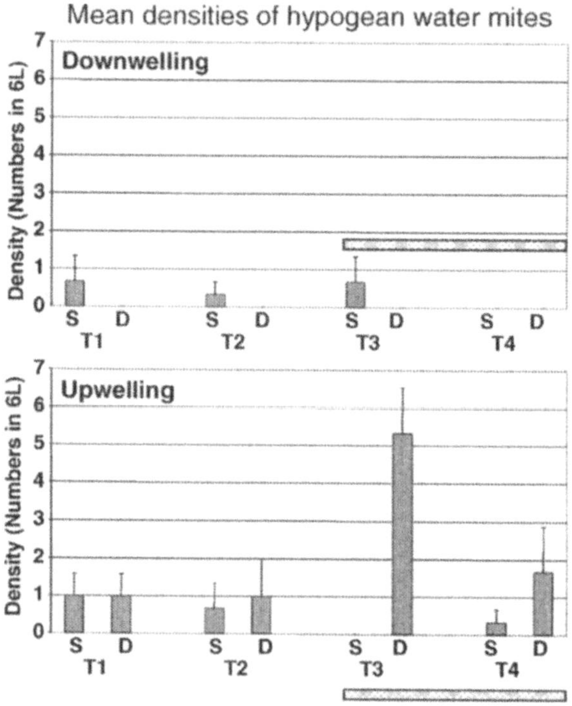

Figure 10. Mean (+ SE) hypogean mite densities (n = 3 wells) at 10 cm depth (S = Shallow) and 50 cm depth (D = Deep) in the second riffle R2 before (T1, T2) and during (T3, T4) an experimental spate at the downwelling (top panel) and upwelling (bottom panel) zones. The hatched bar represents the high flow period.

generally coarse (primarily cobbles and pebbles, ø = −3 to −6), and interstitial flows could possibly be quite high even at 50 cm depth. Boulton and Harvey (2003) suggested that obligate hyporheic mites, especially elongate taxa such as *Wandesia* (*Pseudowandesia*), appear to be poor swimmers and may be prone to being flushed from the interstitial zone by increased hyporheic water velocities. Without autecological data on these species to test this hypothesis, the question remains as to whether spates might play as important a role in structuring hyporheic invertebrate assemblages as they appear to do for surface stream organisms (Lake 2000).

As epigean water mites apparently do not enter the hyporheic zone for refuge from spates in this stream, and most of these mites are weak swimmers that are typically found close to the substratum or among submerged vegetation (Di Sabatino et al. 2000), perhaps catastrophic drift downstream is more common than expected so that many epigean mites are swept away during spates. Although reduced flows among the stems and leaves of submerged waterplants might provide a temporary refuge, especially in backwaters or in

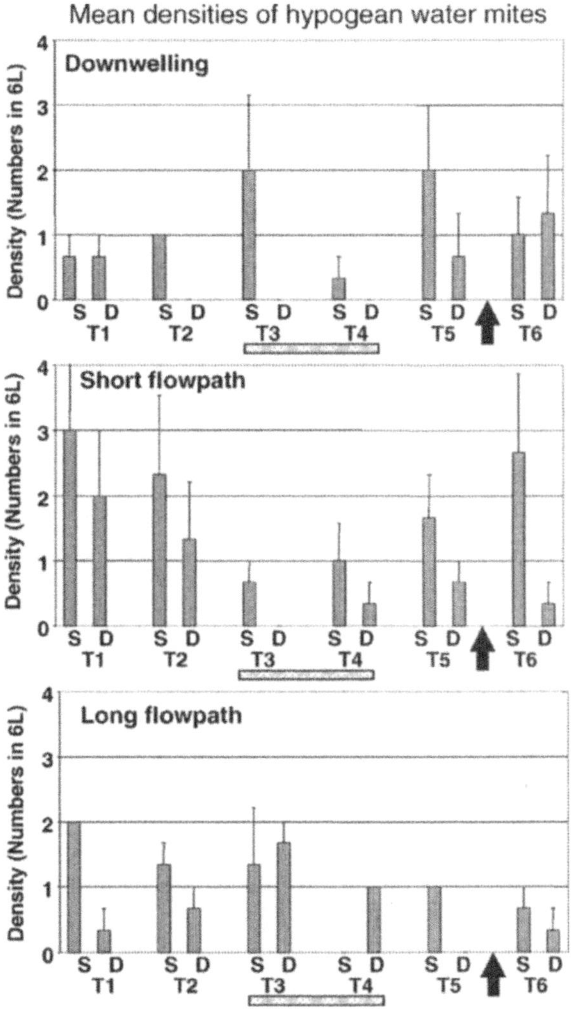

Figure 11. Mean (+1 SE) hypogean mite densities (*n* = 3 wells) at 10 cm depth (S = Shallow) and 50 cm depth (D = Deep) in the first bar B1 before (T1, T2), during (T3, T4) and after the experimental spate (T5) and after a natural spate (T6) in the downwelling zone at the head of the bar (top panel), terminus of the short flow path (middle panel) and terminus of the long flow path (bottom panel). The hatched bar represents the high flow period, the arrow indicates the natural spate.

the lee of rocks and logs, even these refuges would be unavailable during extreme floods (Boulton and Harvey 2003).

This raises the possibility that the experimental and natural spates at each of these habitats were not sufficiently severe to provoke water mites to enter the hyporheic zone in our study. Although our measurements of the strength of

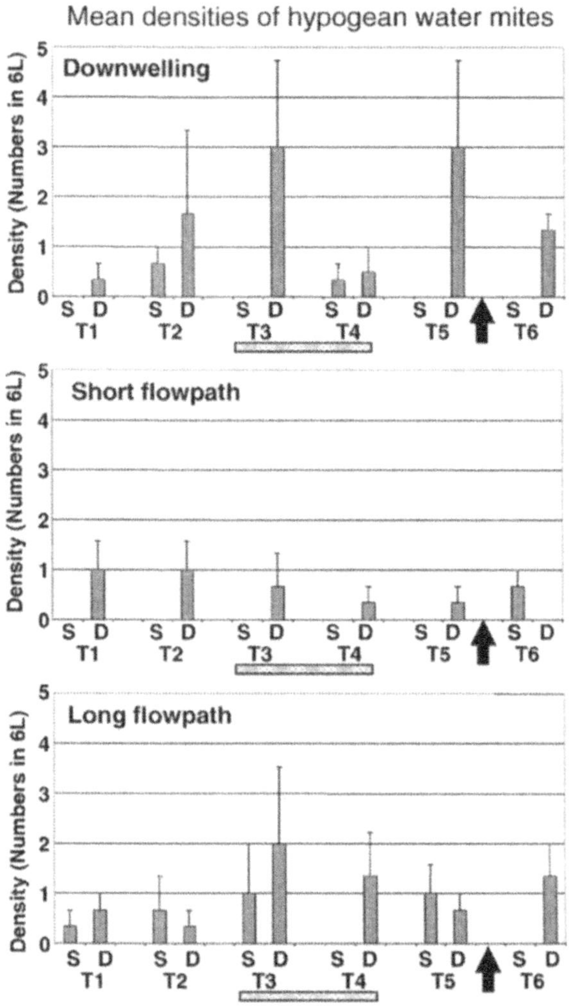

Figure 12. Mean (+1 SE) hypogean mite densities (n = 3 wells) at 10 cm depth (S = Shallow) and 50 cm depth (D = Deep) in the second bar B2 before (T1, T2), during (T3, T4) and after the experimental spate (T5) and after a natural spate (T6) in the downwelling zone at the head of the bar (top panel), terminus of the short flow path (middle panel) and terminus of the long flow path (bottom panel). The hatched bar represents the high flow period, the arrow indicates the natural spate.

hydrological exchange demonstrated that the experimental manipulation did indeed influence the hyporheic zone, only limited movement of sediments occurred in the riffle habitats and none took place at the two lateral bars. Possibly a threshold disturbance strength must be exceeded before water mites or other epigean taxa seek the hyporheic zone as a refuge. Dole-Olivier et al. (1997)

reported that benthic taxa sought refuge in downwelling zones during low and medium floods in a by-passed section of the Rhone River, but at high floods, sediments became unstable and most benthic organisms drifted rather than entered the hyporheic zone. We hypothesise that small spates might induce no effect, perhaps because they are not perceived as disturbances, whereas large spates primarily flush the fauna into the drift. It is possible that the flood refuge hypothesis may hold for intermediate spates in the subtropical river where our work was done. It is also possible that our artificial spates lacked the 'ionic signature' present in natural rain- or melt-water induced spates (Glover and Johnson 1974), and that hyporheic invertebrates require these cues before initiating spate avoidance behaviour.

Hyporheic communities are notoriously patchy and variable (Dole-Olivier and Marmonier 1992; Boulton 2000). In our study, we also found immense fine-scale patchiness and inherent variance in the density data among replicate samples. Even after grouping the water mite data into epigean and hypogean assemblages, the high variance around the means of the triplicate samples often exceeded the variation over time, among sites or between depths. The magnitude of this sampling error implies that more spatial replicates probably need to be taken (Boulton et al. 2003) but this issue must be balanced against the dilemma of artificially depleting the hyporheic zone by repeated sampling from a small area.

In conclusion, the hyporheic refuge hypothesis did not hold for water mites at four sites in this subtropical river, supporting the conclusions from work by Dole-Olivier et al. (1997) and Palmer et al. (1992) indicating limited or no influx by epigean fauna into the sediments in response to spates. Similarly, although hydrological data indicated that the experimental spates altered the strength and sometimes direction of vertical hydraulic gradients, there was no evidence to support the 'wash out' effect observed by Marmonier and Creuzé des Châtelliers (1991). Although our results may be compromised by sampling difficulties and the possibility that the experimental spates were not sufficiently severe to induce a response, our data indicates that the hyporheic refuge hypothesis may not always be an appropriate explanation for rapid post-flood recolonisation. It is likely that some taxa, including water mites, seek alternative refuges from disturbance or are carried downstream as catastrophic drift, and the use of the hyporheic zone as a refuge from floods may be dictated by the strength of the disturbance and substrate composition and stability.

Acknowledgements

We thank Marie-Jo Olivier and Pierre Marmonier for discussion and input into the design of this study and for their immense contribution to the arduous fieldwork, ably supported by Paul Lisle, Trevor Stace and Peter Hancock. Marie-Jo also helped sort many of the samples. We also acknowledge financial support from the Australian Research Council and the CNRS in France.

Perspicacious comments by two anonymous reviewers and Jan Bruin improved the final paper and we thank them.

References

Analytical Software 2000. Statistix 7 User's Manual. Analytical Software, Tallahassee, FL.

Boulton A.J. 1993. Stream ecology and surface-hyporheic hydrologic exchange: implications, techniques and limitations. Aust. J. Mar. Freshwat. Res. 44: 553–564.

Boulton A.J. 2000. The functional role of the hyporheos. Verh. Int. Verein. Limnol. 27: 51–63.

Boulton A.J. and Foster J.G. 1998. Effects of buried leaf litter and vertical hydrologic exchange on hyporheic water chemistry and fauna in a gravel-bed river in northern New South Wales, Australia. Freshwat. Biol. 40: 229–243.

Boulton A.J. and Harvey M.S. 2003. Effects of a simulated spate on water mites in the hyporheic zone of an Australian subtropical river. In: Smith I.M. (ed.), An Acarological Tribute to David R. Cook (From Yankee Springs to Wheeny Creek). Indira Publishing House, Canada, pp. 57–73.

Boulton A.J., Dole-Olivier M.-J. and Marmonier P. 2003. Optimizing a sampling strategy for assessing hyporheic invertebrate biodiversity using the Bou-Rouch method: Within-site replication and sample volume. Arch. Hydrobiol. 156: 431–456.

Boulton A.J., Valett H.M. and Fisher S.G. 1992. Spatial distribution and taxonomic composition of the hyporheos of several Sonoran Desert streams. Arch. Hydrobiol. 125: 37–61.

Brunke M. and Gonser T. 1997. The ecological significance of exchange processes between rivers and groundwater. Freshwat. Biol. 37: 1–33.

Claret C., Marmonier P., Dole-Olivier M.-J., Creuzé des Châtelliers M., Boulton A.J. and Castella E. 1999. A functional classification of interstitial invertebrates: supplementing measures of biodiversity using species traits and habitat affinities. Arch. Hydrobiol. 145: 385–403.

Cook D.R. 1986. Water mites from Australia. Mem. Am. Ent. Inst. 40: 1–568.

Di Sabatino A., Gerecke R. and Martin P. 2000. The biology and ecology of lotic water mites (Hydrachnidia). Freshwat. Biol. 44: 47–62.

Dole-Olivier M.-J. and Marmonier P. 1992. Patch distribution of interstitial communities: prevailing factors. Freshwat. Biol. 27: 177–191.

Dole-Olivier M.-J., Marmonier P. and Beffy J.-L. 1997. Response of invertebrates to lotic disturbance: is the hyporheic zone a patchy refugium? Freshwat. Biol. 37: 257–276.

Gerecke R. and Di Sabatino A. 1996. Water mites (Acari: Hydrachnellae) and spring typology in Sicily. Crunoecia 5: 35–41.

Glover B.J. and Johnson P. 1974. Variations in the natural chemical concentration of river water during flood flows and the lag effect. J. Hydrol. 22: 303–316.

Gordon N.D., McMahon T.A. and Finlayson B.L. 1992. Stream Hydrology: An Introduction for Ecologists. John Wiley and Sons, Chichester.

Harvey M.S. 1990. Pezidae, a new freshwater mite family from Australia (Acarina: Halacaroidea). Invert. Taxon. 3: 771–781.

Harvey M.S. 1998a. Unusual new water mites (Acari: Hydracarina from Australia, Part 1. Rec. West. Aust. Mus. 19: 91–106.

Harvey M.S. 1998b. The Australian Water Mites: A Guide to the Families and Genera. CSIRO Publishing, Melbourne.

Harvey M.S. 2003. *Davecookia*, a new genus of the water mite family Athienemanniidae (Acarina: Hydracarina) from Australia. In: Smith I.M. (ed.), An Acarological Tribute to David R. Cook (From Yankee Springs to Wheeny Creek). Indira Publishing House, Canada, pp. 151–153.

Lake P.S. 2000. Disturbance, patchiness, and diversity in streams. J. N. Am. Benthol. Soc. 19: 573–592.

Marmonier P. and Creuzé des Châtelliers M. 1991. Effects of spates on interstitial assemblages of the upper Rhône River. Importance of spatial heterogeneity. Hydrobiologia 210: 243–251.

Palmer M.A., Bely A.E. and Berg K.E. 1992. Response of invertebrates to lotic disturbance: a test of the hyporheic refuge hypothesis. Oecologia 89: 182–194.

Richards C. and Bacon K.L. 1994. Influence of fine sediment on macroinvertebrate colonization of surface and hyporheic stream substrates. Great Basin Nat. 54: 106–113.

Schwoerbel J. 1959. Ökologishe und tiergeographische Untersuchungen über die Milben (Acari, Hydrachnellae) der Quellen und Bäche des südlichen Schwarzwaldes und seiner Randgebeite. Arch. Hydrobiol. Suppl. 24: 385–546.

Schwoerbel J. 1967. Das hyporheische interstitial als grenzbiotop zwischen oberirdischem und subterranem Ökosystem und seine bedeutung für die primär-evolution von kleinsthöhlenbewohnern. Arch. Hydrobiol. Suppl. 33: 1–62.

Smit H. 1998. A new genus of the water mite family Piersigiidae from Australia (Acari: Hydrachnidia). Rec. West. Aust. Mus. 19: 107–110.

Smith I.M. 1991. Water mites (Acari: Parasitengona: Hydrachnidia) of spring habitats in Canada. Mem. Entomol. Soc. Canada 155: 141–167.

Triska F., Kennedy V.C., Avanzino R.J., Zellweger G.W. and Bencala K.E. 1989. Retention and transport of nutrients in a third-order stream in northwestern California: hyporheic processes. Ecology 70: 1893–1905.

Williams D.D. 1984. The hyporheic zone as a habitat for aquatic insects and associated arthropods. In: Resh V.H. and Rosenberg D.M. (eds), The Ecology of Aquatic Insects. Praeger Publishers, New York, pp. 430–455.

Williams D.D. and Hynes H.B.N. 1974. The occurrence of benthos deep in the substratum of a stream. Freshwat. Biol. 4: 233–256.

Experimental and Applied Acarology **34:** 171–197, 2004.

Environmental parameters determining water mite assemblages in Costa Rica

TOM GOLDSCHMIDT
Universität Karlsruhe, Zoologisches Institut I, Kornblumenstraße 13, D-76128 Karlsruhe, Germany; (e-mail: dc30@rz.uni-karlsruhe.de)

Key words: Biomonitoring, CCA, Ecology, Environmental parameters, Habitat preference, Hydrachnidia, Macroinvertebrates, Neotropics, Springs

Abstract. This paper presents an ecological analysis of the water mite fauna of 350 streams, rivers, springs and standing waters from all main regions, ecosystems and elevations in Costa Rica. From 509 sample sites about 20,000 water mites representing 74 genera in 21 families were collected. 17 habitat types were distinguished and 11 further environmental parameters were measured. The significance of the particular parameters was analysed using canonical correspondence analysis (CCA). Habitat type, elevation, temperature and velocity were identified as major factors determining water mite assemblages. The analyses showed the existence of characteristic water mite coenoses in different neotropical habitats and the suitability of neotropical water mites for biomonitoring programmes.

Introduction

Limnological research on tropical rivers and streams started in the 1960s in the Amazon region (Sioli 1963, 1964, 1965; Fittkau 1964; Gessner 1964) and was mainly focussed on physico-chemical peculiarities and typology of large lowland rivers. During the last decade first attempts were made to relate the invertebrate communities of tropical streams with environmental parameters (Jacobsen et al. 1997; Jacobsen 1998; Jacobsen and Encalada 1998; Thorpe and Lloyd 1999; Miserendino 2001). If these studies took water mites into consideration at all, they were just treated as 'Hydrachnidia' or 'Arachnida'.

However, investigations in North America and Europe found distinct water mite coenoses in different habitats, as well as longitudinal and altitudinal zonations of the water mite fauna (Schwoerbel 1955, 1964; Láska 1957; Young 1969). Schwoerbel (1961) distinguished different spring types and stream zones after characteristic water mite coenoses with typical dominant species and explained the differentiation by physiological adaptations to ecological factors as temperature, current, oxygen saturation, water chemistry and substrate. Young (1969) regarded altitude as main factor for the differentiation of water mite faunas. Angelier et al. (1985) described distinct water mite associations in running waters, correlated with hydrological, topographical and ecological parameters (altitude of source and site, slope, catchment area, water regime type, substrate). The sensitivity of water mites to anthropogenic degradation of aquatic habitats was shown with regard to organic pollution (Schwoerbel 1964; Young 1969; Kowalik and Biesiadka 1982; Cicolani and Di Sabatino 1991; Gerecke and Schwoerbel 1991; Smit and Van der Hammen 1992) as well as structural and hydrological disturbance (Smit and Van der Hammen 1992, 2000; Martin 1996; Van der Hammen and Smit 1996). Whereas strong preferences of several water mite species to particular habitats and environmental parameters are known from the Northern hemisphere (Di Sabatino et al. 2000; Gerecke 2002), the ecology of neotropical water mites is fairly unknown and for most countries of Central and South America rather incomplete species list are the only available information on the water mite fauna (Goldschmidt 2001, 2002). This paper is aimed to enlarge our knowledge on the ecology, distribution and diversity of neotropical water mites and to detect the role of environmental parameters influencing structure and composition of water mite assemblages. The study further provides basic information for the development of biomonitoring programmes.

Material and methods

Study sites

In total 509 sample sites, located in 350 different brooks, streams and rivers, as well as different types of springs and standing waters were studied in Costa Rica (8°00′ to 11°10′N latitude, 82°40′ to 86°00′W longitude). The samples were taken in various regions, ecological zones and elevations all over the country (Figure 1).

The landscape of Costa Rica is richly structured by several mountain chains rising up to 3820 m asl at the Cerro Chirripó in the Cordillera de Talamanca. The annual precipitation varies between an average of about 1500 mm in the north-western dry forest region of Guanacaste up to 7000 mm on the Atlantic slope of the Cordillera de Talamanca (Coen 1983). The rich structured geomorphology of the country, together with the diverse climatic pattern, causes a great diversity of different ecosystems and aquatic habitats.

Water mite sampling

Water mites were collected during 6 months of field work in the dry as well as the wet season in 1995, 1996 and 1997. The samples were taken using a hand-net with a mesh size of 250 μm. In running waters the substratum was stirred by hand, larger stones were turned and moss carpets scratched from stones and boulders, so that the fine detritus together with the dislodged organisms was carried into the net by the current. In standing and slowly flowing waters the net was swept through aquatic plants, tree roots and coarse detritus. In

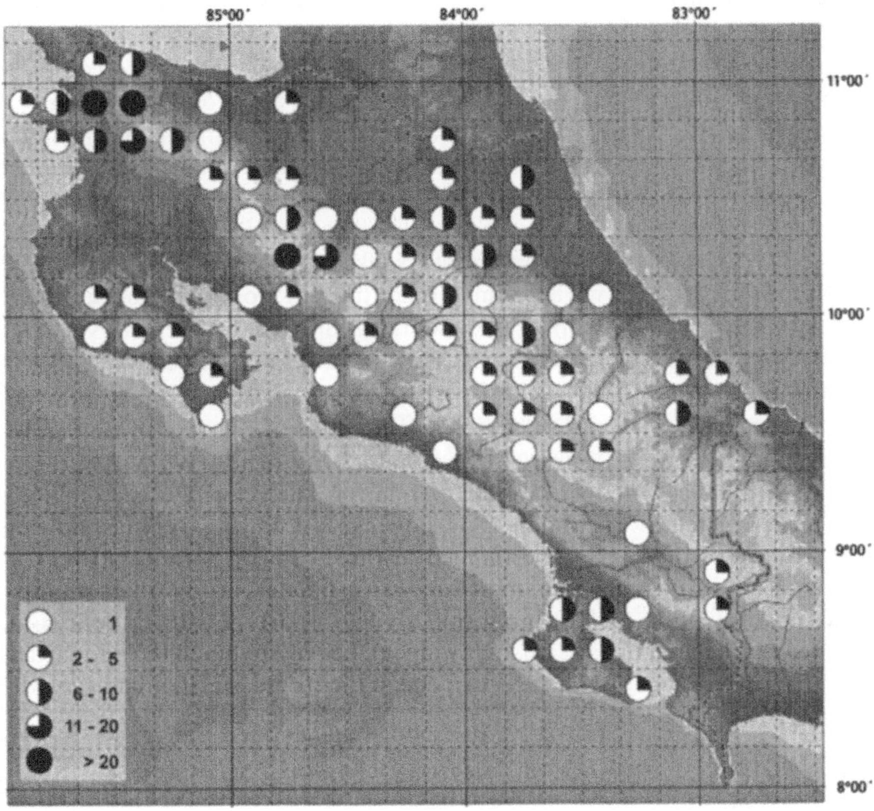

Figure 1. Map of the distribution of all water mite sample sites within Costa Rica. The circles represent the number of sample sites in the respective 10′ × 10′ squares.

seepage areas and small springs with little flow, substratum was gathered in the net by hand. The different microhabitats at every site were collected separately according to their frequency at the site. Samples were taken qualitatively, however with the same intensity at each site to allow comparison among samples. The material collected was washed through a sieve (mesh size approximately 2 mm), transferred to a white tray and sorted in the field. The water mites were picked up individually with eye droppers and fine tweezers and preserved in Koenike's solution (10 parts glycerine : 3 parts acetic acid : 6 parts water). All specimens were identified to genus level using general keys and publications on Central and South American water mites (Cook 1974, 1980, 1988). The genus *Monatractides* was not distinguished from *Torrenticola*. Several genera afterwards were treated in detail and new species were described (Goldschmidt and Gerecke 2003, Goldschmidt 2004, in press). The analyses of the present study were made at genus level to allow including all specimens.

Environmental parameters

Within the sample sites 17 habitat types were distinguished, clustered in three groups (springs, running waters, standing waters) (Table 5): Five spring types (Steinmann 1915; Thienemann 1925; Schwoerbel 1959; Gerecke 1991) and four types of standing waters were differentiated; running waters were classified according to their size, waterfalls, artificial ditches, the hyporheic interstitial and hygropetric areas were regarded separately. Classification of substrates (Table 2) largely follows the system of Braukmann (1987). Four categories of submerged vegetation were distinguished (algae, moss carpets, submerged terrestrial herbs and macrophytes) as well as four types of emergent vegetation (partly flooded terrestrial herbs, reed, swamp vegetation including floating aquatic plants and terrestrial shrubs). The elevation was measured in the field by an altimeter (Altitrek, 0–5000 m asl, Thommen, Switzerland), temperature and conductivity were measured by a conductivity meter (WTW LF 91), pH mainly with indicator paper (Merck, Germany). The velocity was estimated in four relative categories (standing, slow flowing, fast flowing, very fast flowing). The shading of the sites was estimated in three categories (not, partly, completely shaded). Possible contamination was estimated in four classes [unpolluted, possibly polluted (flowing through village, pasture with cattle), slightly polluted (garbage in the stream, many algae), heavily polluted (large quantities of garbage, thick algae or bacteria carpets on the stones, smell, muddy colour)].

Data analysis

Canonical correspondence analysis (CCA) is used to analyse and show relationships between taxa and different environmental variables, especially complex sets of data and large numbers of taxa (Ter Braak 1988; Ruse 1994; Clausen 1998). As first step of a correspondence analysis (CA) a principal component analysis (PCA) of the examined parameters is performed, in order to detect correlation of the parameters within each other and to reduce the large quantity of variables to a limited number of principal components (factors) (Dytham 1999). The first two principal components of the PCA represent the x- and y-axes of the graph of a CCA. In this graph, linear environmental parameters are represented by arrows (vectors) starting in the centre of the diagram. The direction of the vector shows the axis of maximal variability of the parameter, its length corresponds to the importance of the ecological parameter for the variation of the faunal structure along the parameter (Glavac 1996). Nominal parameters are represented by centroids (in the graph symbolised by stars), whose position in the graph corresponds to the weighted average (Ter Braak 1988). The proximity of the centroids (as well as the vectors) to an axis shows the proportion of the respective parameters to the calculation of this axis. The nearness of a sample site or a taxon to a vector (especially its tip) represents the importance of the respective environmental parameter for the site or the taxon. Taxa at the edge of the diagram are often rare and show extreme habitat parameters (Glavac 1996).

To show correlation between the water mite assemblages found in Costa Rica and the environmental parameters, a CCA was performed using the software package CANOCO Version 3.1 (Ter Braak 1990). In this CCA, sample sites were included, when data for all twelve parameters were available and at least two water mite genera were found ($N = 392$ sample sites). The nominal factors habitat type, substrate, submerged and emerged vegetation each were considered separate in three or four classes in the factors-matrix of the CCA. Therefore a total of 23 environmental variables were analysed. Extreme values of taxa were excluded: ten genera only found in single specimens as well as two mass-findings of *Hygrobates* and *Sperchon*. Abundance data were log $x + 1$ transformed prior to the CCA to normalise differences between sites and to avoid overrating high and underrating low abundance.

To explore the importance of the habitat type for the distribution of the Costa Rican water mite fauna, a 'simple' CA was carried out, only considering genera and habitat types. As CA is more sensitive than CCA to rare taxa and samples with few taxa (Glavac 1996), in this analysis only genera represented by more than three specimens and habitat types examined more than five times were taken into consideration. This reduced the data matrix for CA to 59 genera and eleven habitat types (helocrenes, rheocrenes, rheopsammocrenes, spring brooks, brooks, streams, rivers, waterfalls, residual pools of intermittent streams, lakes and swamps). The black dots in the CA graph indicate locations of highest abundance or highest probability of particular genera (Glavac 1996). Genera are clustered according to their relationship to the habitat types, represented by grey dots. The CA was performed using the programme NTSYS 1.6 (Applied Biostatistics Inc., Setauket New York). Extreme values of taxa were excluded (see above), abundance data of the individual genera were log $x + 1$ transformed prior to the analysis.

Results

Environmental parameters

A summary of the values of all continuous parameters is given in Table 1.

The water mite samples were taken between 5 and 3560 m asl, unevenly distributed over the different elevations: 52% of all sample sites lay below 500 m, only 6.4% above 2000 m asl. However, the altitudinal distribution of the sample sites generally reflects the proportions of the particular elevations in the total surface of Costa Rica (Figure 2): In total 63% of the country is below 500 m, elevations above 2000 m are only found in the summit regions of the Cordillera Central and the ridge of the Cordillera de Talamanca (in total 6.5% of the country's surface). In the present study elevations between 0 and 100 m are slightly under-represented (21% of all samples compared to 35% of the

Table 1. Overview of the continuous environmental parameters measured at the sample sites of Costa Rican water mites.

Parameter	Min	Max	Median	Mean	SD
Altitude (m asl)	5	3560	345	676.5	766.0
Temperature (°C)	7.1	37.3	23.5	22.3	4.7
Velocity	0	3	1	1.4	0.8
Depth (cm)	0	100	20	22.6	13.9
Shading	0	2	1	1.2	0.7
Conductivity ($\mu S\ cm^{-1}$)	7	3000	101	153.3	207
pH	4.0	8.9	7.0	7.0	0.8
Contamination	0	3	0	0.3	0.7

Min = minimum, Max = maximum, SD = standard deviation.

country's surface), elevations above 3000 m are slightly over-represented (3.2% compared to 0.5%) (Figure 2).

The mean temperature of the sample sites from sea level to the high mountains went down by 4.8 °C per 1000 m altitude. The highest temperature (37.3 °C) was measured in a residual pool at 5 m asl, the lowest (7.1 °C) in a brook at 3340 m asl (Table 1). The most abundant relative velocity classes were 2 (fast flowing, 43.4%) and 1 (slow flowing, 37.3%), followed by 0 (standing waters, 13.4%) and 3 (very fast flowing, 5.9%). Most samples (79.4%) were taken in depths between 10 and 40 cm, one sample was taken at

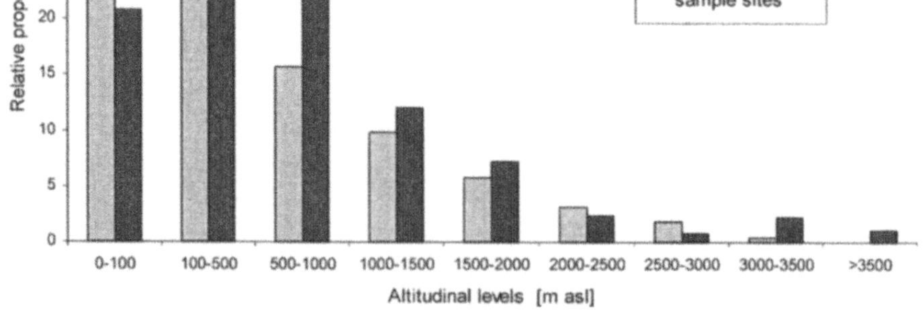

Figure 2. Relative proportions of the sample sites in the particular altitudinal levels with regard to all sample sites in Costa Rica and relative proportions of the particular altitudinal levels with regard to the whole surface of the country. As the areas above 3500 m asl represent only 0.05% of the surface of Costa Rica, their proportion is not visible in the diagram.

100 cm depth, however generally habitats deeper than 50 cm were sampled only at the bank. Of all sample sites 20% were not shaded (0), 45% partly shaded (1) and 35% completely shaded (2). At most sample sites conductivity values were measured between 7 and 180 $\mu S\ cm^{-1}$, at five sites conductivity was above 1000 $\mu S\ cm^{-1}$, with a maximum of 3000 $\mu S\ cm^{-1}$. At this latter site (Río Murcielago, northwest Costa Rica) the typical genera of slow flowing lowland streams (*Koenikea*, *Unionicola*, *Hygrobates*, *Limnesia*) were found. At most sample sites pH values were measured in a neutral range between 6.0 and 8.0 (Table 1). The lowest values were 4.0 in four sample sites in streams influenced by volcanic activity, the highest values between 8.5 and 9.0 were found in residual pools of drying out intermittent streams. At pH 4.0 exclusively *Sperchon* and *Corticacarus* were found, both in high abundances. Most sites were undisturbed by anthropogenic contamination, however near towns and villages, running waters were sometimes heavily polluted by sewage and garbage; 9% of all sites in running waters were slightly or heavily polluted. In heavily polluted streams, mainly *Sperchon* was found regularly, however also *Atractides*, *Atractidella* and *Hygrobates* were found.

At all sample sites mineral substrates dominated (40% in total), followed by living plants (18%) and dead organic matter (12%). In 30% of all samples, the substrate was classified as a mixture of different types (Table 2).

In running waters, mineral substrates were most abundant (45%), in springs and standing waters, mixed substrates (52 and 54%) dominated (Figure 3).

Table 2. Relative abundance of the different substrates at all sample sites.

	Type of substrate	Relative abundance [%]
Mineral substrates	Psammal (sand)	0.6
	Akal (gravel)	4.4
	Mesolithal (stones)	17.5
	Macrolithal (boulders)	2.8
	Mixed mineral substrates	14.9
Dead organic matter	Leaf package	0.2
	Micropelal (fine particles)	0.4
	Macropelal (coarse particles)	8.2
	Mixed organic substrates	2.8
Living plants	Lithophytal	11.3
	Phytal	1.0
	Terrestrial plants	3.4
	Tree roots	1.2
	Floating macrophytes	0.8
	Various plants	0.6
Mixed substrate classes	Min.-org. mixed substrates	18.7
	Min.-pl. mixed substrates	3.8
	Org.-pl. mixed substrates	3.6
	All types mixed	3.6

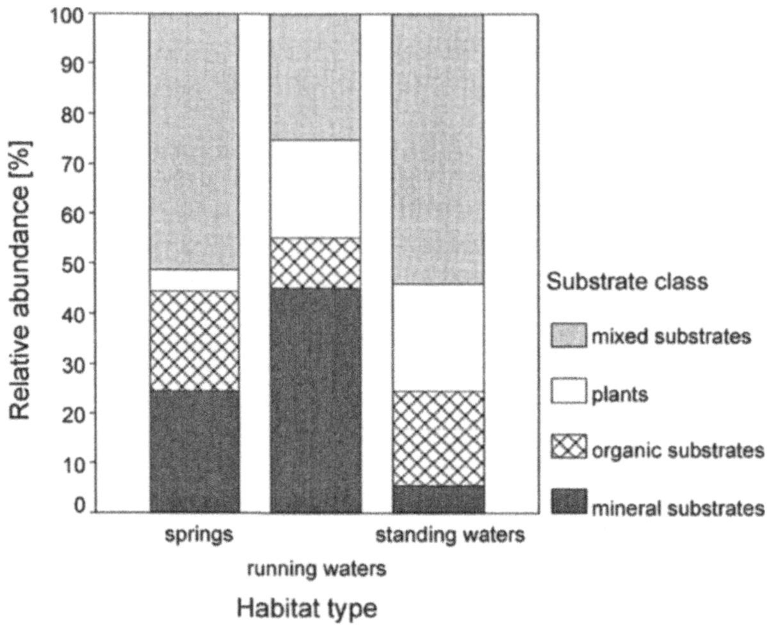

Figure 3. Proportions of the substrate classes in the different habitat types.

Water mite communities

A total of 19,443 specimens of water mites were collected, representing 74 genera in 21 families (Table 3). Some genera were found in high steadiness and high numbers of specimens, the four most frequent genera (*Torrenticola*, *Atractides*, *Hygrobates* and *Sperchon*) were found at more than 100 sample sites each. The six most abundant genera (each found in more than 1000 individuals) represent 80% of all specimens. However, most genera were present only at few sample sites (19 genera were found at one single site each) and in low abundance (10 genera were represented by single findings only).

The four axes of the CCA together explain 66% of the variance of the genera with regard to the considered parameters (Table 4). The *x*-axis of the CCA is mainly dominated by the physico-chemical parameters, with the highest correlation between relative velocity, temperature and altitude (Figure 4). The *y*-axis mainly depends upon the habitat type. Therefore temperature, altitude, velocity and habitat type (especially springs) are the most important parameters for the grouping of the sample sites (Figure 4) and the distribution of the water mite genera (Figure 5). A strong negative correlation exists between temperature and altitude, visible in the exactly opposed vectors (Figure 4). Water depth is negatively correlated with springs, reflecting the fact, that most of the investigated springs are very shallow, mainly built up by seepage sand

Table 3. Number of specimens and sample sites of all water mite genera collected in Costa Rica.

	Specimens	Sample sites
Hydrachnoidea		
Hydrachnidae		
Hydrachna	4	3
Limnocharidae		
Limnochares (Limno)	15	5
Rhyncholimnochares (Rhy)	92	34
Eylaioidea		
Eylaidae		
Eylais	2	1
Hydryphantoidea		
Hydryphantidae		
Almuerzothyas	2	1
Eupatrella (Eup)	128	6
Hydryphantes	6	1
Neocalonyx (Neocal)	23	2
Protzia (Protz)	73	24
Trichothyas	1	1
Hydrodromidae		
Hydrodroma (Hyd)	228	36
Rhynchohydracaridae		
Clathrosperchon (Cla)	118	4
Rhynchohydracarus	2	2
Lebertioidea		
Sperchontidae		
Sperchon (Spe)	2571	108
Anisitsiellidae		
Bandakia	3	1
Fuenticola	2	1
Mamersellides	7	2
Mamersopsides	1	1
Nilotonia (Nil)	34	6
Lebertiidae		
Estelloxus (Est)	12	9
Lebertia (Leb)	64	17
Oxidae		
Frontipoda (Fro)	23	8
Oxus (Oxu)	7	3
Torrenticolidae		
Neoatractides (Neoatr)	259	65
Pseudotorrenticola (Pse)	34	11
Torrenticola (Tor)	3706	222
Hygrobatoidea		
Limnesiidae		
Centrolimnesia (Cen)	20	5
Guanacastacarus	10	1
Limnesia (Limne)	1693	76
Meramecia	1	1
Neomamersa (Neomam)	8	5
Neotorrenticola (Neotor)	145	29
Neotyrrellia (Neotyr)	16	8
Protolimnesia (Pro)	164	11
Rheolimnesia (Rhe)	7	6
Tyrrellia	5	3
non det.	13	
Hygrobatoidea		
Omartacaridae		
Omartacarus (Oma)	7	3
Hygrobatidae		
Atractidella	286	66
Atractides (Atr)	919	171
Aspidiobates (Asp)	4	2
Corticacarus (Cor)	1797	67
Hygrobatella	4	2
Hygrobates (Hyg)	4471	143
Mapuchacarus (Map)	6	3
Megapella (Meg)	15	6
Paraschizobates (Par)	47	30
Stylohygrobates	1	1
Maharashtracarus	1	1
non det.	22	
Unionicolidae		
Koenikea (Koe)	1157	79
Neumania (Neu)	139	39
Schadeella (Scha)	6	1
Unionicola (Uni)	226	41
non det.	45	
Pionidae		
Piona	11	4
Pionopsis (Pio)	8	1
non det.	11	
Aturidae		
Albaxona	2	1
Albia (Alb)	10	7
Aturus (Atu)	25	8
Axonopsalbia (Axonopsa)	9	6
Axonopsella (Axonopse)	4	2
Axonopsis (Axonopsi)	73	20
Diamphidaxona (Dia)	8	5
Frontipodopsis	1	1
Javalbia (Jav)	3	2
Kongsbergia (Kon)	4	3
Notaxona	1	1
Pseudoaxonopsalbia	1	1
Szalayella	1	1
non det.	12	
Arrenuroidea		
Mideopsidae		
Mideopsis (Mid)	56	19
Krendowskiidae		
Geayia (Gea)	59	16
Krendowskia (Kre)	34	3
Roqueella	1	1
non det.	2	
Hungarohydracaridae		
Cubanohydracarus (Cub)	5	2
Arrenuridae		
Arrenurus (Arr)	353	45
Stygarrenurus (Sty)	92	13
non det.	5	
Total	19443	365

The abbreviations used in the figures 5, 6, 7 are given in brackets.

Table 4. Basic data of the CCA (CANOCO 3.12-output).

**** Summary ****					
Axes	1	2	3	4	Total inertia
Eigenvalues	.527	.263	.229	.149	11.256
Species–environment correlations	.876	.720	.751	.661	
Cumulative percentage variance					
of species data	4.7	7.0	9.1	10.4	
of species–environment relation	29.6	44.4	57.3	65.7	
Sum of all unconstrained eigenvalues					11.256
Sum of all canonical eigenvalues					1.779

and gravel deposits. Velocity is negatively correlated with temperature and positively with altitude and submersed vegetation of the classes one and two (algae and moss) (Figure 4).

Most sample sites form a cloud of dots extended along the *x*-axis (Figure 4). The left part, along the vectors of velocity and altitude, includes the fast flowing lotic habitats and the samples from elevations between 2000 and 3500 m asl. On the right side are mainly slow flowing streams and rivers, at the right edge of the diagram standing waters. Residual pools of intermittent streams partly group with other lentic habitats, partly they are mixed with slow flowing brooks in the lower right part of the diagram. A single black dot at the lower right edge of the cloud represents a brook with a conductivity of 3000 $\mu S\ cm^{-1}$. In the upper part of the diagram, positively correlated with the *y*-axis, springs form a remote group. A rheocrene spring at 3450 m is isolated at the left upper edge and a faunistically very particular helocrene at the right upper edge. Two isolated dots between the springs and the large group of running waters represent waterfalls (Figure 4).

The genera, similar to the sample sites (Figure 4), are mainly clustered along the *x*-axis of the CCA-diagram (Figure 5). An isolated group far on the positive range of the *y*-axis consists of spring-dwelling genera (among others *Nilotonia*, *Eupatrella* and *Stygarrenurus*). In the left part of the main group, along the vectors of altitude and velocity, there is a cluster of rheobiotic genera with *Aturus*, *Lebertia*, *Corticacarus* and others. Closer to the centre, a group of more euryoecious rheophilous genera such as *Torrenticola* and *Atractides* is found. Further genera (*Centrolimnesia*, *Arrenurus*, *Neumania*, *Unionicola*, *Limnesia*, *Koenikea*), in their higher positive correlation with the *x*-axis, reflect decreasing preference for cold, high-altitude, lotic habitats and increasing occurrence in warm, low-altitude, lentic habitats (Figure 5).

Genera (and sample sites) were weighted in the analysis, according to their abundance (Ter Braak 1988). The highest weighting has the genus *Torrenticola*, followed by *Hygrobates*, *Atractides*, *Sperchon*, *Koenikea*, *Limnesia* and *Corticacarus*. Genera (and samples) at the edge of the diagram generally possess lower values and have lower influence on the analysis.

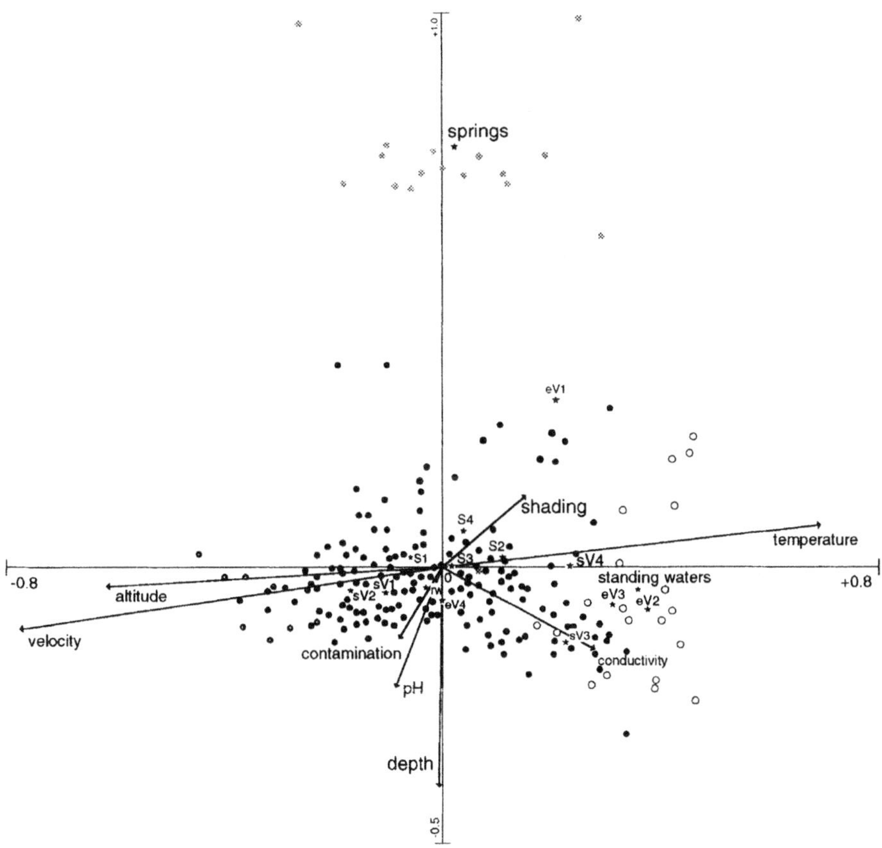

Figure 4. CCA-biplot of the water mite sample sites and the investigated environmental parameters. Stars (centroids) represent the habitat types, substrates as well as emergent and submersed vegetation. rw = running waters; S1 = mineral substrates, S2 = organic substrates, S3 = plants, S4 = mixed substrates; eV1 = terrestrial herbs, eV2 = reed, eV3 = swamp plants, swimming leaves, eV4 = terrestrial shrubs, sV1 = algae, little moss, sV2 = moss, sV3 = terrestrial herbs, sV4 = macrophytes. Sample sites in running waters are represented by black dots, in springs by grey dots and in standing waters by white dots. Black rings represent running waters above 3000 m asl.

Habitat type

Most samples (83%) were taken in running waters (with 47% brooks dominated by far) and 87% of all water mite specimens were collected there (Table 5). Of the 74 genera found, 64 occurred in running waters (32 exclusively there), 32 (six exclusively) in springs and 22 (four exclusively) in standing waters. Twelve genera were found in all three habitat types. The mean number of genera per sample was highest in running waters (4.0), the average number of specimens per sample site was highest in residual pools of intermittent streams (69.2) (Table 5).

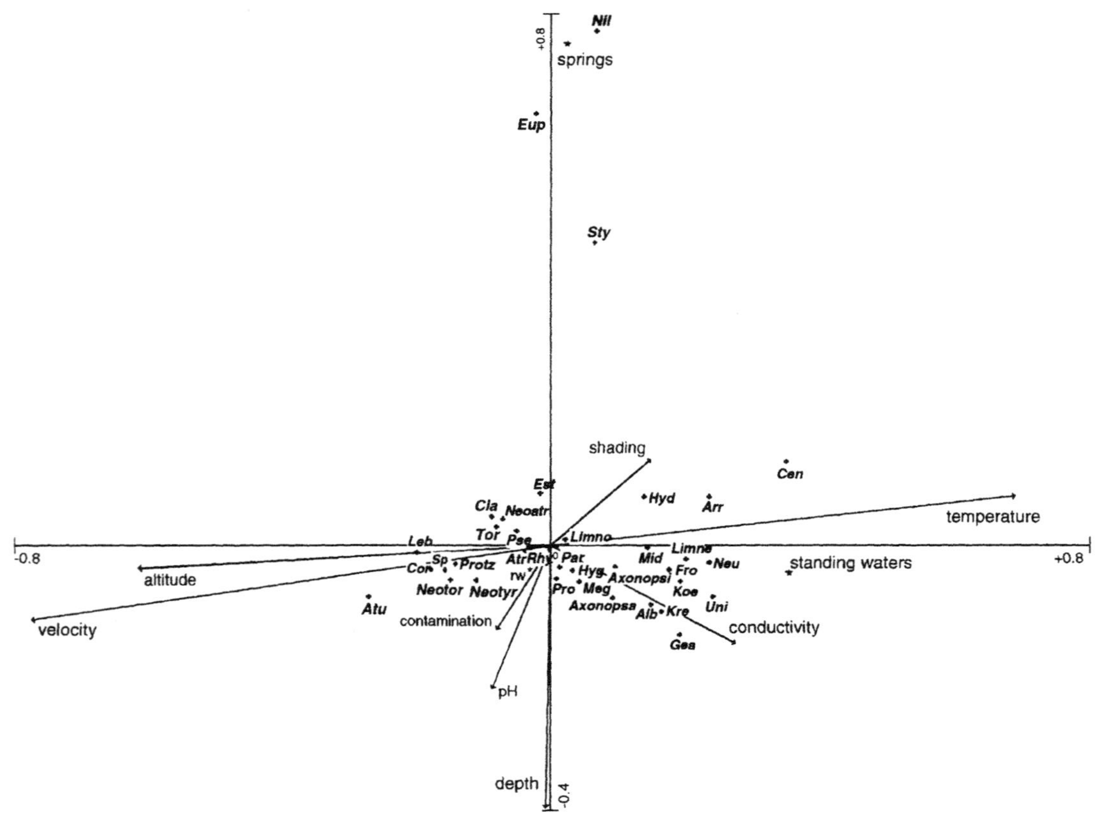

Figure 5. CCA-biplot of the water mite genera, the linear environmental parameters and the habitat types. Crosses represent the water mite genera, see Table 3 for key to abbreviations.

Table 5. Abundance and proportion of the 17 particular habitat types within all sample sites in Costa Rica.

Habitat types	Samples	Proportion [%]	Genera	Specimens	Genera/sample		Specimens/sample
					Mean	Min-max	
Springs, total	50	9.8	32	692	2.3	1–9	13.8
Helocrenes	7	1.4	14	128	2.4	1–7	18.3
Rheocrenes	17	3.3	18	158	2.2	1–9	9.3
Rheohelocrenes	6	1.2	6	21	1.7	1–3	3.5
Rheopsammocrenes	16	3.1	14	356	2.9	1–7	22.3
Limnocrenes	4	0.8	4	29	1.5	1–2	7.3
Running waters, total	421	82.8	64	16882	4.0	1–16	40.1
Spring brooks (< 1 m)	48	9.4	40	1226	3.8	1–11	25.5
Brooks (1–5 m)	238	46.8	50	11382	3.8	1–13	47.8
Streams (5–10 m)	87	17.1	39	3345	4.9	1–16	38.5
Rivers (> 10 m)	32	6.3	15	800	3.5	1–7	25.0
Ditches	1	0.2	3	7	3.0	3–3	7.0
Waterfalls	7	1.4	8	17	1.4	1–2	2.4
Hygropetric zones	4	0.8	6	33	1.8	1–3	8.3
Hyporheic interstitial	4	0.8	12	72	3.8	2–5	18.0
Standing waters, total	38	7.6	22	1760	3.2	1–8	46.3
Residual pools of intermittent streams	13	2.6	11	900	3.0	1–6	69.2
Lakes	10	2.0	13	303	2.4	1–5	30.3
Swamps	12	2.4	16	550	4.2	1–8	45.8
Periodic swamps	3	0.6	6	7	3.0	1–5	2.3

Number of genera and specimens found in the particular habitat types and average number per sample.

A 'simple' CA, performed in order to analyse the faunistic differences of the habitat types, shows clear differentiation (Figure 6). Brooks, streams and rivers form a dense cluster, with spring brooks at the edge of this group; waterfalls are far isolated. Swamps and lakes form a separate group. Residual pools of intermittent streams fall in between lotic habitats. Helocrenes are isolated from other springs (Figure 6).

A group of rheophilic genera (group I) that includes *Sperchon*, *Corticacarus*, *Atractides*, *Lebertia*, *Neotorrenticola*, *Neocalonyx*, *Protolimnesia* and *Limnochares*, is closely associated with the lotic habitat types. *Aturus* and *Neotyrrellia* form a group of rheophilic genera also found in waterfalls (group II). Group III is composed of genera that occur mainly in running waters, but are also found in standing waters, mainly the residual pools of intermittent streams. Typical genera of running waters, such as *Hygrobates*, *Axonopsis* and *Geayia* belong to this group, as well as the genera *Koenikea* and *Unionicola* that are also found in many other habitat types. Group IV (*Hydrodroma*, *Arrenurus* and *Limnesia*) likewise colonises lentic and lotic habitats, but also occurs in different types of springs. Group V combines the crenobiotic genera

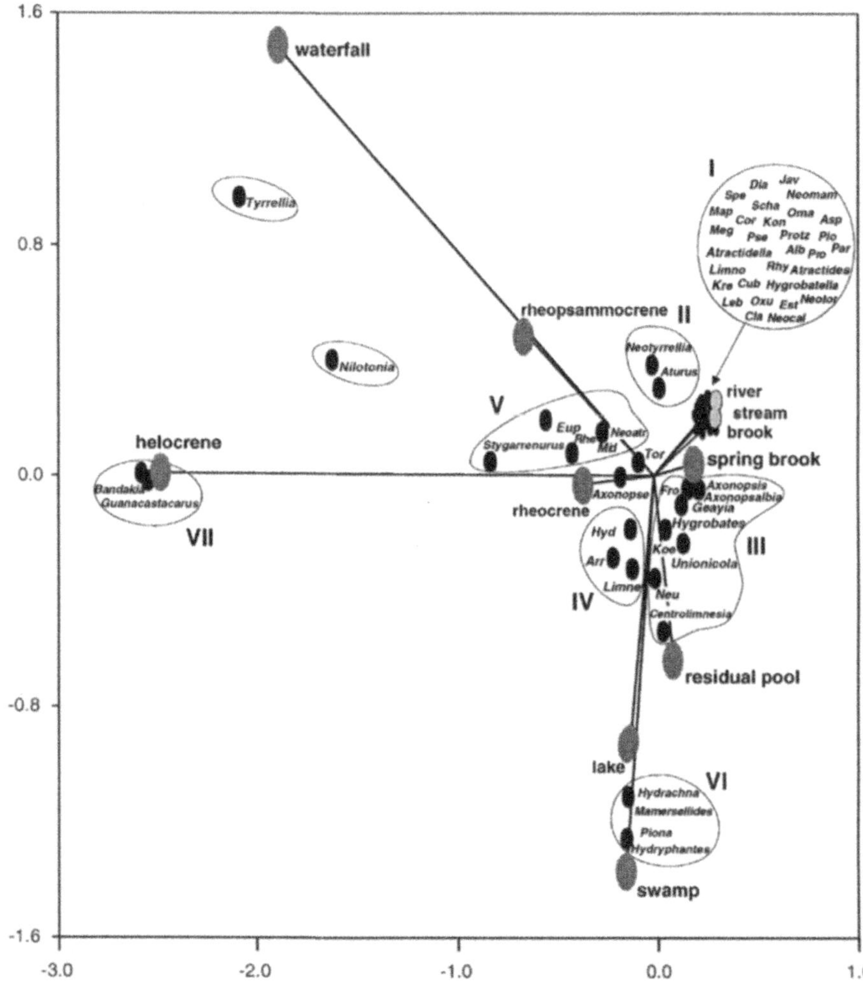

Figure 6. CA-plot of the habitat types and water mite genera. Grey ovals represent the habitat types, black ovals represent the position of the genera in the ordination created by the CA. The groups of genera formed relatively to their correlation with the habitat types, are labelled I–VII (see text). See Table 3 for key to abbreviations.

Stygarrenurus, *Eupatrella* and *Rheolimnesia* with the rheobiotic-crenophilous *Mideopsis* and *Neoatractides*. The genera *Mamersellides*, *Hydrachna*, *Piona* and *Hydryphantes* form the exclusively limnophilous group VI. The genera *Bandakia* and *Guanacastacarus* represent the strictly helocrenobiotic group VII. The crenobiotic genus *Nilotonia* is found in various types of springs and in waterfalls, *Tyrrellia* is mainly found in waterfalls.

For better resolution of the lotic groups, the habitats sampled more intensively (spring brooks, brooks, streams and rivers), are displayed in a separate CA-plot (Figure 7). Most genera of the groups I–V of Figure 6 are included.

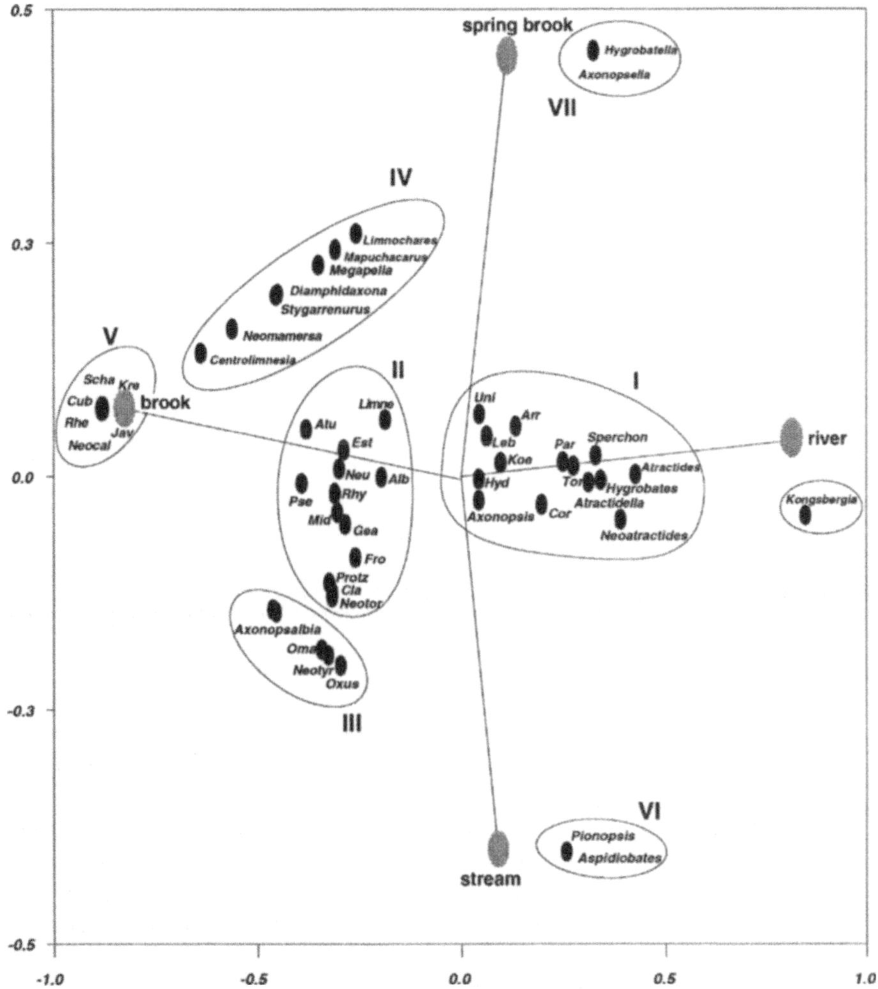

Figure 7. CA-plot of the running waters and the lotic water mite genera. See Table 3 for key to abbreviations (for explanation see Figure 6.).

A large group (I) of typical rheophilous genera (*Sperchon*, *Torrenticola*, *Atractides*, *Lebertia*, etc.) were found in all four habitat types and hence are not associated with a single type. The genera of group II (among others *Limnesia*, *Aturus* and *Protzia*) were mainly found in brooks and streams, but also in spring brooks. Group III was found in brooks and streams; group IV in spring brooks and brooks. The genera of group V (*Rheolimnesia*, *Neocalonyx* and others) are restricted to brooks, the genera of group VI (*Pionopsis*, *Aspidiobates*) to streams. *Hygrobatella* and *Axonopsella* (group VII) are associated with spring brooks. *Kongsbergia* is the only genus strongly associated with rivers, however this habitat preference is verified by only two samples.

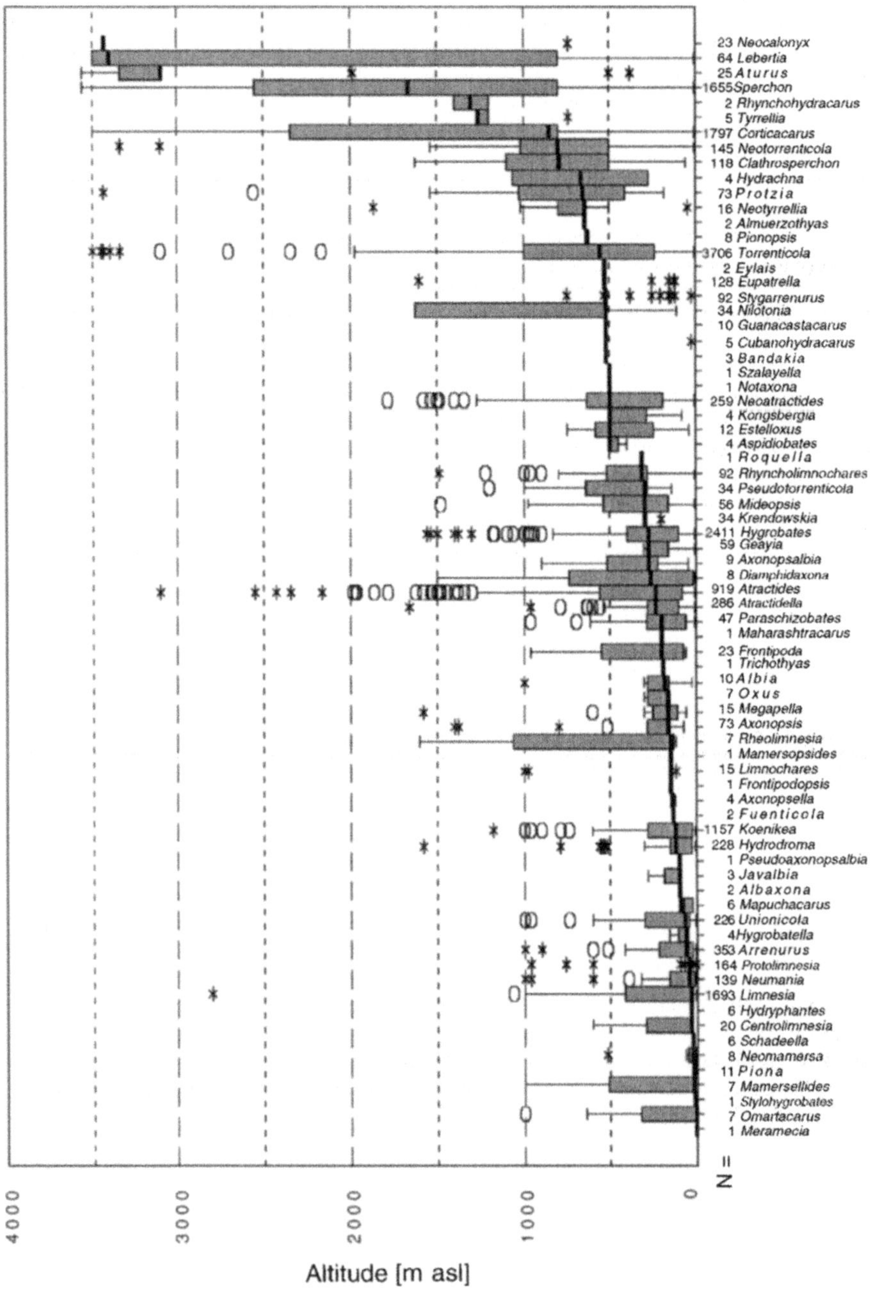
23 Neocalonyx
64 Lebertia
25 Aturus
1655 Sperchon
2 Rhynchohydracarus
5 Tyrrellia
1797 Corticacarus
145 Neotorrenticola
118 Clathrosperchon
4 Hydrachna
73 Protzia
16 Neotyrrellia
2 Almuerzothyas
8 Pionopsis
3706 Torrenticola
2 Eylais
128 Eupatrella
92 Stygarrenurus
34 Nilotonia
10 Guanacastacarus
5 Cubanohydracarus
3 Bandakia
1 Szalayella
1 Notaxona
259 Neoatractides
4 Kongsbergia
12 Estelloxus
4 Aspidiobates
1 Roquella
92 Rhyncholimnochares
34 Pseudotorrenticola
56 Mideopsis
34 Krendowskia
2411 Hygrobates
59 Geayia
9 Axonopsalbia
8 Diamphidaxona
919 Atractides
286 Atractidella
47 Paraschizobates
1 Maharashtracarus
23 Frontipoda
1 Trichothyas
10 Albia
7 Oxus
15 Megapella
73 Axonopsis
7 Rheolimnesia
1 Mamersopsides
15 Limnochares
1 Frontipodopsis
4 Axonopsella
2 Fuenticola
1157 Koenikea
228 Hydrodroma
1 Pseudoaxonopsalbia
3 Javalbia
2 Albaxona
6 Mapuchacarus
226 Unionicola
4 Hygrobatella
353 Arrenurus
164 Protolimnesia
139 Neumania
1693 Limnesia
6 Hydryphantes
20 Centrolimnesia
6 Schadeella
8 Neomamersa
11 Piona
7 Mamersellides
1 Stylohygrobates
7 Omartacarus
1 Meramecia
N =
4000
3000
2000
1000
0
Altitude [m asl]

Altitude

Most of the abundant genera were found over a wide altitudinal range (Figure 8), e.g. *Torrenticola* and *Atractides* were present from 5 to 3500 m, respectively, 3100 m asl. However, *Lebertia*, a rare genus with only 64 specimens from 17 sample sites, also showed a wide altitudinal range. Some of these genera (e.g. *Torrenticola*) were equally represented at all elevations, whereas the abundance of others (e.g. *Lebertia*) was strongly biased towards higher or lower elevations. Of all sample sites of *Lebertia*, 56% were below 1000 m, 17% above 3000 m, however only 27% of all specimens were found below 1000 m, 53% above 3000 m asl. Due to these differences, the sample data of the genera were weighted by the number of specimens. Therefore the median of their occurrence can be shifted towards the higher or lower end of their altitudinal range. In *Lebertia* (sample sites from 10 to 3500 m asl) the weighted median is at 3400 m asl, as most specimens were found at high elevations (the unweighted median of all sample sites would have been at 720 m). *Sperchon* was found from 10 to 3560 m asl, the weighted median is at 1660 m asl (unweighted at 900 m), whereas the weighted median of *Limnesia* (samples sites from 5 to 2800 m asl) is at 30 m asl (unweighted at 240 m) (Figure 8).

The altitudinal distribution of the genera shows a grouping of the weighted medians in distinct levels (Figure 8). Several genera are characteristic for particular altitudinal levels. Most of the genera (46 of 74) have a clear centre of distribution (shown by the weighted medians) between 5 and 310 m asl. Within the more abundant genera, *Limnesia*, *Neumania*, *Protolimnesia*, *Arrenurus* and *Unionicola* represent a lowland-group with a main distribution below 100 m asl. *Hydrodroma* and *Koenikea* represent the transition to the next altitudinal level (100–310 m asl) which is characterised by the genera *Axonopsis*, *Paraschizobates*, *Atractidella*, *Atractides*, *Geayia*, *Hygrobates* and *Rhyncholimnochares* (Figure 8). Three further groups follow with weighted medians at 500–560 m asl (14 genera, characterised mainly by *Neoatractides* and *Torrenticola*), 630–665 m asl (five genera, with *Protzia* as characteristic genus) and 790–850 m asl (*Clathrosperchon*, *Neotorrenticola*, *Corticacarus*). For three genera the weighted median lies at 1260–1660 m asl (*Sperchon* is the characteristic genus). In three high mountain genera (*Lebertia*, *Aturus*, *Neocalonyx*) the median of their distribution is at 3100–3430 m asl. Even as these latter

Figure 8. Altitudinal distribution of all findings of all water mite genera collected in Costa Rica. In the 'Box and Whisker-Plots', the grey 'Box' represents the interquartil-range, i.e. all values between the 25. and 75. percentil; the black line marks the median. The vertical lines ('Whiskers') connected to the box to the top and the bottom, cover values up to the 1.5-fold of the interquartil-range. Values beyond this range (runaways) are represented by circles; values beyond the threefold of the interquartil-range (extreme values) are represented by stars. Findings were weighted by the number of specimens prior to the analysis (the median in the boxplots can therefore be shifted to the edge of the interquartil ranges). Two mass findings (916 specimens of the genus *Sperchon*, 2060 specimens of the genus *Hygrobates*) were excluded from the analysis. *N* – number of specimens.

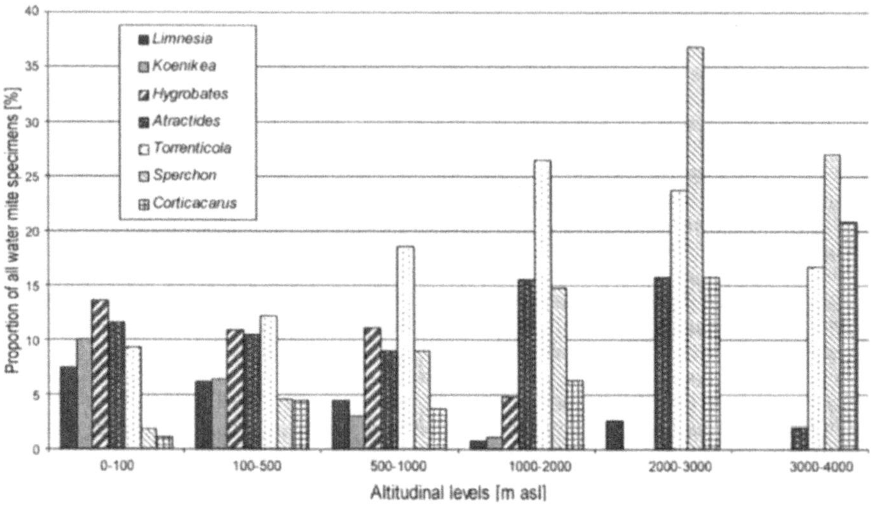

Figure 9. Presence of the seven most abundant Costa Rican water mite genera at particular altitudinal levels.

genera were also found at low elevations, 53–91% of the specimens were collected above 3000 m asl (Figure 8).

The seven most abundant genera *Limnesia*, *Koenikea*, *Hygrobates*, *Atractides*, *Torrenticola*, *Sperchon* and *Corticacarus*, in total 82% of all specimens), alternated in their dominance in the different altitudinal levels: *Hygrobates* dominated in the lowland, *Torrenticola* was the most abundant genus at mid elevations, *Sperchon* dominated at the higher elevations above 2000 m asl (Figure 9). Whereas several genera showed similar abundance at lower elevations, a few genera clearly dominated the fauna at higher elevations. Most rare genera were limited to lower elevations; no genus only found at a single sample site was found above 1000 m, only ten genera are found above 2000 m asl.

Temperature

For the particular genera the weighted medians of the temperature of the sample sites ranged between 10.0 and 30.2 °C. Most samples were taken between 20 and 30 °C (Figure 10).

Torrenticola was found over the complete range of temperature, and another four abundant genera in their distribution showed a range of more than 20 °C (Figure 10). Only six genera (*Arrenurus*, *Neumania*, *Koenikea*, *Limnesia*, *Hygrobates* and *Torrenticola*) were found in habitats with temperatures above 35 °C. At the low end of the range, *Lebertia*, *Aturus* and *Neocalonyx* had a median temperature of 10–13 °C. Elevation and temperature preferences show

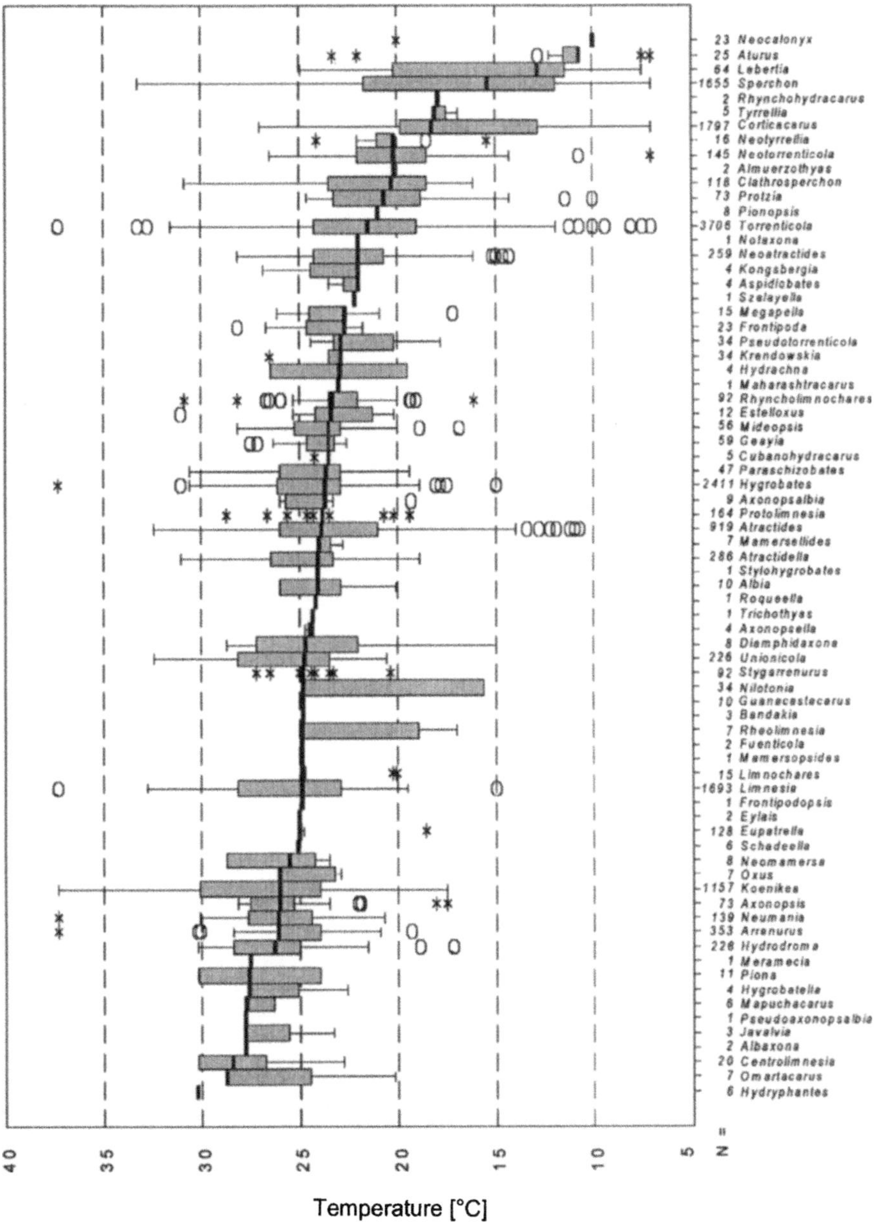

Figure 10. Water temperatures at the sample sites of all water mite genera collected in Costa Rica. *N* – number of specimens. (For further explanations see Figure 8.)

a negative relationship, with low-altitude taxa tending to be in warm water and vice versa; however the correlation is not perfect, and the rank order of genera in Figures 8 and 10 are not identical.

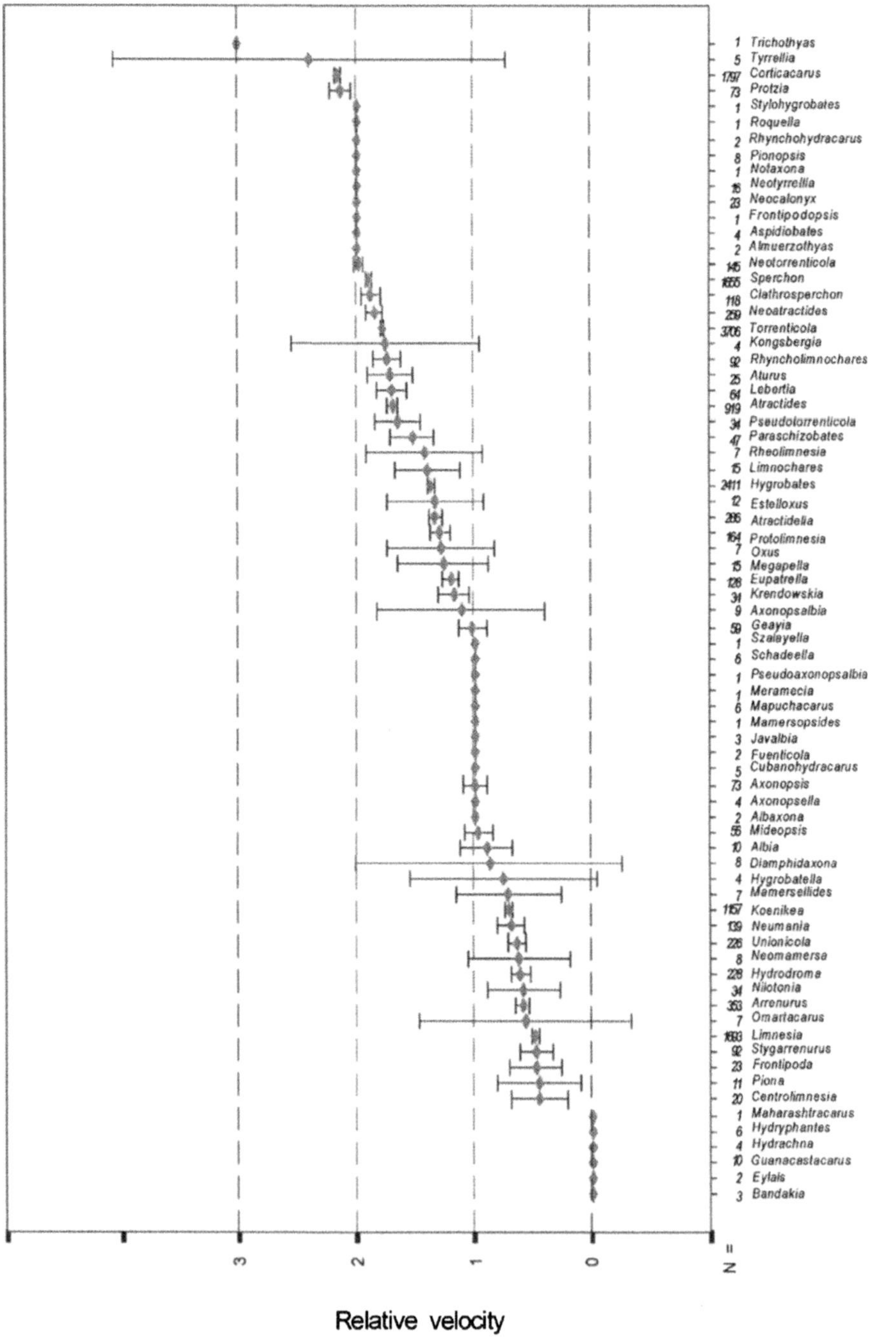

Figure 11. Velocity preferences of the Costa Rican water mite genera. The average values are shown, weighted by the number of specimens, and their 95%-confidence interval. *N* – number of specimens.

Velocity

Most genera of Costa Rican water mites were found at sample sites with different stream velocities. The most abundant genera (more than 100 sample sites) were found at all four classes of velocity. Even typical inhabitants of running waters, such as *Torrenticola*, *Sperchon* or *Corticacarus*, in single findings were also found in standing waters such as residual pools of intermittent streams. However the mean values (weighted by the number of specimens) of the 74 genera collected in Costa Rica reflect clear differences in their rheophily (Figure 11).

Limnesia, *Arrenurus*, *Hydrodroma* and *Koenikea* together with several other genera can be classified as limno- to slightly rheophilous, the mean values of the relative velocity of their sample sites were 0.5–1 (0-standing, 1-slowly flowing). *Torrenticola*, *Neoatractides* and *Sperchon* characterise a group of clearly rheophilous genera with mean values at relative velocities between 1.5 and 2 (fast flowing). The highest values of velocity preferences within the more abundant genera showed *Protzia* (mean value 2.1) and *Corticacarus* (mean value 2.2) (Figure 11).

The six most abundant genera clearly differ with regard to their velocity preferences (Figure 12). 58% of all specimens of *Limnesia* were found in standing water, *Koenikea* was mainly found in slow flowing habitats (60% of all specimens). The genera *Hygrobates*, *Torrenticola*, *Sperchon* and *Corticacarus* were found in habitats of all velocities. However, *Hygrobates* showed a preference for slow flowing habitats, with 53% of all specimens collected in this type of running waters. Whereas more than 70% of the specimens of *Torrenticola*, *Sperchon* and *Corticacarus* were found in fast flowing habitats. The strongest rheophily revealed *Corticacarus*, only 0.1% of all specimens of this genus were found in standing waters, 23% were found in very fast flowing habitats (Figure 12).

Discussion

Multivariate analysis

In a CCA faunistical data as well as environmental parameters were found to differentiate the sample sites of Costa Rican water mites. The analysis of the significance of the particular environmental parameters proved velocity, altitude, temperature and habitat type, especially springs, to be the dominating factors for the composition of the water mite assemblages. In the analysis displayed by the CCA, springs are far isolated from all other habitat types. The samples from waterfalls are also separated, especially in a simple CA only based upon the water mite communities, these habitats are clearly isolated. The results of the CCA correspond to an investigation on the diversity of benthic macroinvertebrates in Ecuadorian streams (Jacobsen et al. 1997). The authors

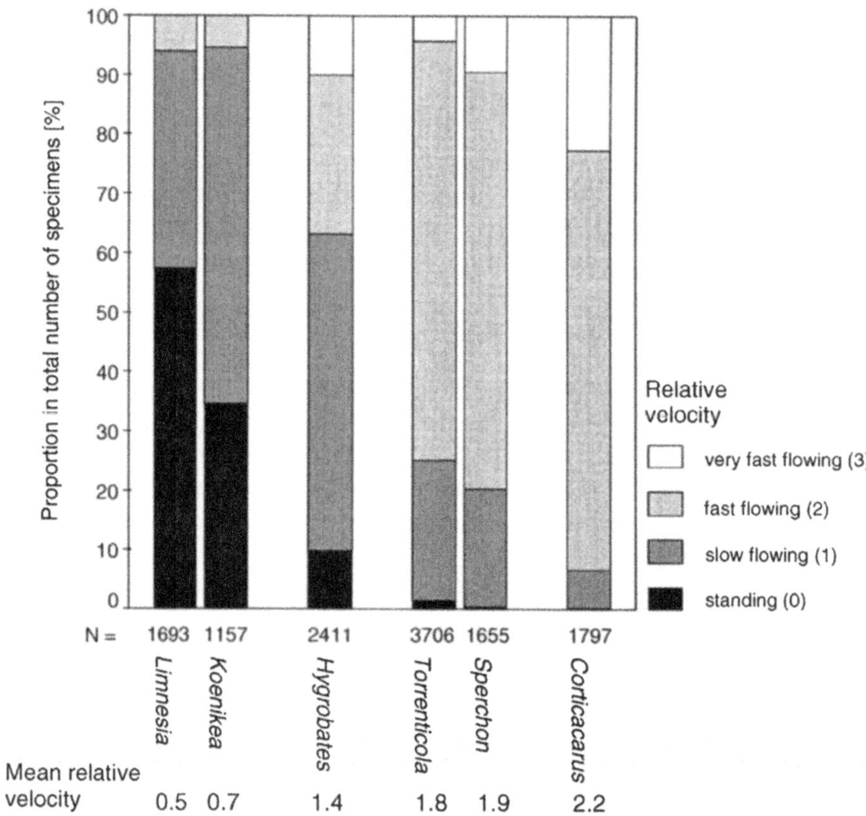

Figure 12. Distribution of the specimens of the most abundant genera among four velocity classes. *N* – number of specimens.

regard temperature as the main factor for faunistic variation between different streams. In a CCA of the macroinvertebrate communities of running waters on the Caribbean island of St. Lucia, Thorpe and Lloyd (1999) found the type of land use as well as the substrate composition as main factors to explain the variability of the communities. In a study on Brazilian streams, Melo and Froehlich (2001) regard the physical structure, especially stream width and sediment structure, as the main factors that determine the macroinvertebrate communities.

Multivariate analyses of water mite assemblages in France revealed a 'locality complex' of mainly topographical and hydrological parameters (altitude, water regime, slope, substrate etc.) correlated with different water mite associations in running waters (Angelier et al. 1985). In a CCA of the water mite fauna of coastal regions in the Netherlands, Smit and Van der Hammen (1992) found water chemistry (pH, nutrients), emergent vegetation and dimension of the habitat as most important variables for the water mite fauna.

Habitat preferences and environmental parameters

With regard to the habitat preference and ecology of Costa Rican water mites – especially the more abundant genera – it must be considered, that the analyses presented here are based upon genus level. Most of the euryoecious genera probably encompass an unknown number of more stenoecious species, which in total cover a wide range of habitats. *Torrenticola* for example was found in 15 of the 17 studied habitat types, *Arrenurus*, *Hygrobates* and *Koenikea* in 12 each, *Limnesia* in 11. In an analysis of lotic water mites at species level, Angelier et al. (1985) also could apply some ecological assignations to whole genera. Some habitat preferences of Costa Rican water mites are already visible at genus level (e.g. six genera were exclusively found in springs).

Young (1969) found distinct water mite faunas in different habitat types in North America and considered altitude as the main factor. In several studies in Central Europe, Schwoerbel (1955, 1961, 1964) regarded temperature (besides current, substrate, pollution) as the principal factor for the differentiation of water mite coenoses in streams. In temperate latitudes the presence of particular microhabitats has also great significance for the composition of the water mite fauna (Martin 1996; Di Sabatino et al. 2000). Within particular habitat types and elevations in Costa Rica clear differences are visible in the composition of the substrates, however substrate showed no strong influence on the water mite community.

The application of the concept of biocoenotic longitudinal zonation (Illies and Botosaneanu 1963; Illies 1964) to tropical running waters demonstrates a fundamental problem: On the one hand, there is a clear zonation along individual streams, in which several parameters, such as temperature, width, current and sediment structure change at the same time. On the other hand, there are differences in between streams of the same type (stream order, width, substrate) at different elevations and latitudes; which are mainly explained by differences in temperature (Jacobsen et al. 1997). Ramírez and Pringle (2001) describe a faunal change along several streams in Costa Rica at the transition from the foothills of the Cordillera Central to the Caribbean lowland. Within these streams mainly the sediment structure changed from predominant coarse- to fine-sediment. In the present study no difference was made between sample sites in various streams at different elevations and sample sites along an altitudinal gradient within the same stream.

Water chemistry showed minor influence on the differentiation of Costa Rican water mites, however, conductivity seems to have some influence. Schwoerbel (1961) emphasised the importance of limestone on the occurrence of water mite species in Central Europe. In two mountain streams in Guatemala, mainly varying in their conductivity, Böttger (1980) found clearly different water mite assemblages. Tropical running waters, especially in the lowland, due to the extreme lack in nutrients of most tropical soils and the efficiency of the nutrient-recycling of the surrounding forests, in general show very low conductivity (Fittkau 1982; Payne 1986; Umaña et al. 1999). In Costa

Rica soils are younger and mainly built up by volcanic rock rich in minerals, therefore in most areas conductivity is not as low as it is in many other tropical regions. The very high conductivity of some Costa Rican streams is caused by three geological and hydrological reasons: Geo-chemical influences due to recent volcanism (Pringle 1991; Umaña et al. 1999), accumulation of ions as a result of drying out of intermittent streams and influx of sea-water by the tide. However, for further discussion on the influence of chemical parameters, analysis on species level would be necessary.

Conclusions

In this study, habitat type, elevation, temperature and velocity clearly turned out to be the major factors that influence Costa Rican water mite assemblages. The sediment structure showed less influence in the analysis than would have been expected from other studies. Maybe further investigations should also emphasise on this aspect. Further more detailed analysis with regard to different microhabitats, contamination and geological peculiarities of the water chemistry will be possible with the data at hand, as soon as a larger part of the material is determined to species level. Even though clear habitat preferences are already obvious at genus level, a much more differentiated picture can be expected at species level.

Several studies in Europe and North America already proved water mites as excellent indicators for water quality (Schwoerbel 1964; Young 1969; Kowalik and Biesiadka 1982; Cicolani and Di Sabatino 1991; Gerecke and Schwoerbel 1991; Smit and Van der Hammen 1992), natural habitat structure and hydrology (Smit and Van der Hammen 1992, 2000; Martin 1996; Van der Hammen and Smit 1996). The present study showed that the biological foundations for the construction of freshwater biomonitoring programmes on the basis of water mite assemblages also exist in the neotropics. Already at the present state of knowledge, specific water mite coenoses in particular habitats and distinct factors determining the structure of the water mite fauna are obvious. The use of biomonitoring programmes to survey the water quality is of special importance in the tropics, as besides the advantage of the long temporal and multifactorial integration of biological monitoring methods, expensive chemical laboratory analyses can be economised. However for the establishment of biomonitoring programmes more information on the taxonomy of neotropical water mites as well as their ecology at species level would be necessary. That point has to be emphasised, especially with regard to the still high demand and poor funding of taxonomic research.

Acknowledgements

The author is deeply indebted to Reinhard Gerecke (Tübingen, Germany) for his constant help and advice during the work on the Costa Rican water mites.

I would like to thank my former colleagues at the Department of Limnology at the University of Münster (Germany), especially Norbert Kaschek and my friends in the Area de Conservación Guanacaste (Costa Rica), especially Maria Marta Chavarria Diaz and Roger Blanco, for their help and encouragement during the field work in the ACG. I also thank Helga Bültmann (Münster) for her great help with the CCA and Antonio Di Sabatino (L'Aquila, Italy) for his great help with the CA. The field work in Costa Rica was supported by grants from Boeringer Ingelheim Fonds and DAAD. I thank Heather Proctor (Edmonton, Canada) and Antonio Di Sabatino for revising and improving this paper.

References

Angelier E., Angelier M.-L. and Lauga J. 1985. Recherches sur l'écologie des hydracariens (Hydrachnellae, Acari) dans les eaux courantes. Annales de Limnologie 21(1): 25–64.

Böttger K. 1980. Qualitative und quantitative Benthonstudien an Bergbächen in Guatemala (Zentralamerika) unter besonderer Berücksichtigung der Hydrachnellae (Acari). Archiv für Hydrobiologie. 88: 96–119.

Braukmann U. 1987. Zoozönologische und saprobiologische Beiträge zu einer allgemeinen regionalen Bachtypologie. Ergebnisse der Limnologie (Archiv für Hydrobiologie, Beiheft) 26: 1–355.

Cicolani B. and Di Sabatino A. 1991. Sensitivity of water mites to water pollution. In: Dusbábek F. and Bukva V. (eds), Modern Acarology, Vol. 1. Academia, Prague, SPB Academic Publishing bv, The Hague, pp. 465–474.

Clausen S.-E. 1998. Applied Correspondence Analysis – An Introduction. Sage Publ., Thousand Oaks, pp. 1–69.

Coen E. 1983. Climate. In: Janzen D.H. (ed.), Costa Rican Natural History. University of Chicago Press, London, pp. 35–46.

Cook D.R. 1974. Water mite genera and subgenera. Memoirs of the American Entomological Institute 21: 1–860, I–VII.

Cook D.R. 1980. Studies on Neotropical water mites. Memoirs of the American Entomological Institute 31: 1–645.

Cook D.R. 1988. Water mites from Chile. Memoirs of the American Entomological Institute 42: 1–356.

Di Sabatino A., Gerecke R. and Martin P. 2000. The biology and ecology of lotic water mites (Hydrachnidia). Freshwater Biology 44: 47–62.

Dytham C. 1999. Choosing and Using Statistics: A Biologists's Guide, Vol. 1. Blackwell Science Ltd., Oxford, p. 218.

Fittkau E.-J. 1964. Remarks on limnology of Central-Amazon rain-forest streams. Verhandlungen der Internationalen Vereinigung für theoretische und angewandte Limnologie 15: 1092–1096.

Fittkau E.-J. 1982. Struktur, Funktion und Diversität zentralamazonischer Ökosysteme. Archiv für Hydrobiologie. 95: 29–45.

Gerecke R. 1991. Taxonomische, faunistische und ökologische Untersuchungen an Wassermilben (Acari, Actinedida) aus Sizilien unter Berücksichtigung anderer aquatischer Invertebraten. Lauterbornia 7: 1–303.

Gerecke R. 2002. The water mites (Acari, Hydrachnidia) of a little disturbed forest stream in southwest Germany – a study on seasonality and habitat preference, with remarks on diversity patterns in different geographical areas. In: Bernini F., Nannelli R., Nuzzaci G. and de Lillo E. (eds), Acarid Phylogeny and Evolution: Adaptation in Mites and Ticks. Proceedings of the 4th Symposium of the EURAAC. Kluwer Academic Publishers, Dordrecht, pp. 69–90.

Gerecke R. and Schwoerbel J. 1991. Water quality and water mites (Acari, Actinedida) in the upper Danube region, 1959–1984. In: Dusbábek F. and Bukva V. (eds), Modern Acarology Vol. 1. Academia, Prague, and SPB Academic Publishing bv, The Hague, pp. 483–491.

Gessner F. 1964. The limnology of tropical rivers. Verhandlungen der Internationalen Vereinigung für theoretische und angewandte Limnologie 15: 1090–1091.

Glavac V. 1996. Vegetationsökologie: Grundfragen, Aufgaben, Methoden. Gustav Fischer Verlag, Jena, Stuttgart, Lübeck, Ulm, pp. 1–358.

Goldschmidt T. 2001. Die Wassermilbenfauna Costa Ricas (Hydrachnidia, Acari) – Systematik, Ökologie, Diversität und Zoogeographie. Dissertation. Westfälische Wilhelms-Universität, Münster pp. 1–425.

Goldschmidt T. 2002. The biodiversity of neotropical water mites. In: Bernini F., Nannelli R., Nuzzaci G. and de Lillo E. (eds), Acarid Phylogeny and Evolution: Adaptation in Mites and Ticks. Proceedings of the 4th Symposium of the EURAAC. Kluwer Academic Publishers, Dordrecht, pp. 91–99.

Goldschmidt T. 2004. Untersuchungen über Wassermilben der Familie Anisitsiellidae Koenik 1910 aus der Neotropis (Acari, Actinedida, Hydrachnidia). Senkenbergiana Biol. 83: 2. in press.

Goldschmidt T. 2004. Studies on neotropical Limnesiidae Thor 1900. Part I: Neomamersinae Lundblad, 1953. Part II: Tyrrelliinae Koenike, 1910 Part III: Protolimnesiinae Viets, 1940. Archiv für Hydrobiologie, Supplement 151: 1–2, Monographische Studien, pp. 1–24, 25–68, 69–123.

Goldschmidt T. and Gerecke R. 2003. Studies on hydryphantid water mites (Acari: Actinedida: Hydrachnidia) from Central and South America. In: Smith I.M. (ed.), An Acarological Tribute to David R. Cook – From Yankee Springs to Wheeny Creek. Indira Publishing House, West Bloomfield, Michigan, pp. 83–150.

Illies J. 1964. The invertebrate fauna of the Huallaga, a Peruvian tributary of the Amazon River, from the sources down to Tingo Maria. Verhandlungen der Internationalen Vereinigung für theoretische und angewandte Limnologie 15: 1077–1083.

Illies J. and Botosaneanu L. 1963. Problèmes et méthods de la classification et de la zonation écologique des eaux courantes, considerées surtout du point de vue faunistique. Mitteilungen der Internationalen Vereinigung für theoretische und angewandte Limnologie 12: 1–59.

Jacobsen D. 1998. The effect of organic pollution on the macroinvertebrate fauna of Ecuadorian highland streams. Archiv für Hydrobiologie 143: 179–195.

Jacobsen D. and Encalada A. 1998. The macroinvertebrate fauna of Ecuadorian highland streams in the wet and dry season. Archiv für Hydrobiologie 142: 53–70.

Jacobsen D., Schultz R. and Encalada A. 1997. Structure and diversity of stream invertebrate assemblages: the influence of temperature with altitude and latitude. Freshwater Biology 38: 247–261.

Kowalik W. and Biesiadka E. 1982. Occurence of water mites (Hydracarina) in the River Wieprz polluted with domestic-industry sewage. Acta Hydrobiologica 23(4): 331–348.

Láska F. 1957. Beitrag zur Lösung einiger ökologischer Probleme bei torrenticolen Wassermilben aus dem Reichensteiner Gebirge in Schlesien. Abhandlungen des naturwissenschaftlichen Vereins zu Bremen 35(1): 63–74.

Martin P. 1996. Faunistisch-ökologische Benthosstudien an den Wassermilben (Hydrachnidia, Acari) zweier Bäche des norddeutschen Tieflandes (Ostholsteinisches Hügelland, Schleswig-Holstein). Faunistisch-Ökologische Mitteilungen 7: 153–167.

Melo A.S. and Froehlich C.G. 2001. Macroinvertebrates in neotropical streams: richness patterns along a catchment and assemblage structure between two seasons. Journal of the North American Benthological Society 20: 1–16.

Miserendino M.L. 2001. Macroinvertebrate assemblages in Andean Patagonian rivers and streams: environmental relationships. Hydrobiologia 444: 147–158.

Payne A.I. 1986. The Ecology of Tropical Lakes and Rivers. John Wiley and Sons, Chichester, pp. 1–301, I–VIII.

Pringle C.M. 1991. Geothermally modified waters surface at La Selva biological station, Costa Rica: volcanic processes introduce chemical discontinuities into lowland tropical streams. Biotropica 23: 523–529.

Ramirez A. and Pringle C.M. 2001. Spatial and temporal patterns of invertebrate drift in streams draining a Neotropical landscape. Freshwater Biology 46: 47–62.

Ruse L.P. 1994. Chironomid microdistribution in gravel of an English chalk river. Freshwater Biology 32: 533–550.

Schwoerbel J. 1955. Ökologische Studien an torrentikolen Wassermilben (Hydrachnellae, Acari). Ein Beitrag zur Ökologie unserer Schwarzwaldbäche. Archiv für Hydrobiologie, Supplement 22(3–4): 530–537.

Schwoerbel J. 1959. Ökologische und tiergeographische Untersuchungen über die Milben (Acari, Hydrachnellae) der Quellen und Bäche des südlichen Schwarzwaldes. Archiv für Hydrobiologie, Supplement 24: 385–546.

Schwoerbel J. 1961. Die Bedeutung der Wassermilben für die biozönotische Gliederung. Verhandlungen der Internationalen Vereinigung für theoretische und angewandte Limnologie 14: 355–361.

Schwoerbel J. 1964. Die Wassermilben (Hydrachnellae und Limnohalacaridae) als Indikatoren einer biozönotischen Gliederung von Breg und Brigach sowie der obersten Donau. Archiv für Hydrobiologie, Supplement 27 1(4): 386–417.

Sioli H. 1963. Beiträge zur regionalen Limnologie des Amazonasgebietes. V. Die Gewässer der Karbonstreifen Unteramazoniens (sowie einige Angaben über Gewässer der anschließenden Devonstreifen). Archiv für Hydrobiologie 59: 311–350.

Sioli H. 1964. General features of the limnology of Amazonia. Verhandlungen der Internationalen Vereinigung für theoretische und angewandte Limnologie 15: 1053–1058.

Sioli H. 1965. Bemerkungen zur Typologie amazonischer Flüsse. Amazoniana 1: 79–83.

Smit H. and Van der Hammen H. 1992. Water mites as indicators of natural aquatic ecosystems of the costal dunes of the Netherlands and Northwestern France. Hydrobiologia 231: 51–64.

Smit H. and Van der Hammen H. 2000. Atlas van de Nederlandse Watermijten (Acari: Hydrachnidia). Nederlandse Faunistische Mededelingen 13: 1–272.

Steinmann P. 1915. Praktikum der Süßwasserbiologie. Band 1: Die Organismen des fließenden Wassers. Gebrüder Borntraeger, Berlin, pp. 1–184, I–VI.

Ter Braak C.J.F. 1988. CANOCO – a FORTRAN program for canonical community ordination by [partial] [detrended] [canonical] correspondence analysis, principal components analysis and redundancy analysis (version 2.1). Agricultural Mathematics Group, Wageningen, Technical Report LWA-88-02.

Ter Braak C.J.F. 1990. Update Notes: Canoco Version 3.10. Agricultural Mathematics Group, Wageningen, pp. 1–35.

Thienemann A. 1925. Die Binnengewässer Mitteleuropas. Eine limnologische Einführung. Verlag Schweizerbart, Stuttgart, pp. 1–255.

Thorpe T. and Lloyd B. 1999. The macroinvertebrate fauna of St. Lucia elucidated by canonical correspondance analysis. Hydrobiologia 400: 195–203.

Umaña G., Haberyan K.A. and Horn S.P. 1999. Limnology in Costa Rica. In: Wetzel R.G. and Gopal B. (eds), Limnology in Developing Countries Vol. 2. International Scientific Publishers, New Delhi, pp. 33–62.

Van der Hammen H. and Smit H. 1996. The water mites (Acari: Hydrachnidia) of streams in the Netherlands: Distribution and ecological aspects on a regional scale. Netherlands Journal of Aquatic Ecology 30(2–3): 175–185.

Young W.C. 1969. Ecological Distribution of Hydracarina in North Central Colorado. The American Midland Naturalist 82: 367–401.

Experimental and Applied Acarology **34:** 199–210, 2004.

Diversity, distribution and ecology of water mites (Acari: Hydrachnidia and Halacaridae) in high Alpine lakes (Central Alps, Italy)

A. DI SABATINO[1,*], A. BOGGERO[2], F.P. MICCOLI[1] and B. CICOLANI[1]

[1]*Dipartimento di Scienze Ambientali, University of L'Aquila, Via Vetoio 20, I-67100 Coppito L'Aquila, Italy;* [2]*CNR-Institute for Ecosystem Study, Largo Tonolli, 52 - 28922 Verbania, Italy;* **Author for correspondence (e-mail: adisab@univaq.it)*

Key words: Biodiversity, Distribution, Ecology, High Alpine lakes, Hydrochemistry, Multivariate analysis, Water mites

Abstract. Information on water mite assemblages from high elevation lentic biotopes is scant. A survey of 14 small Alpine lakes located between1900 and 2400 m a.s.l. in Italy resulted in the discovery of 17 species of Hydrachnidia and a single species of freshwater Halacaridae. *Arrenurus conicus* and *Lebertia tuberosa* were the most widespread and abundant species; *Lebertia sefvei, Lebertia rufipes, Oxus setosus, Panisus torrenticolus* and *Sperchon glandulosus* were also widely distributed but relatively less abundant. *Atractides fissus* and *Arrenurus conicus* are recorded for the first time from Italy. In contrast to mid/low elevation lakes and ponds, water mite assemblages of alpine lakes are less diverse and are composed mainly of rheo- and crenobiontic taxa, most of which are cold-stenothermic. Typical standing water dwellers represented only a small fraction (23%) of the species sampled. A principal component analysis conducted on lake environmental variables resulted in a clear separation of the lakes mainly based on ionic contents, pH and temperature. Water mites seem to be less influenced by these factors than by temperature fluctuations and habitat stability and heterogeneity. We conclude with some considerations of the influence of abiotic and biotic factors on the altitudinal and latitudinal distribution pattern of water mites.

Introduction

Compared to terrestrial and marine habitats, studies on spatial patterns of biodiversity in freshwaters are scant and limited to a few taxa (Heino 2002; Wetzel 2002). Recently, Smit et al. (2000) demonstrated the importance of small lentic habitats to explain water mite colonization patterns and the role of these habitats as refugia for rare and endemic species. However, faunistic and ecological characteristics of water mite assemblages from high mountain lentic biotopes are still poorly investigated.

Information extracted from a large electronic database which includes all published records of water mites in Mediterranean countries (Di Sabatino and Gerecke, unpublished) show that only 56 records matched the query 'lentic habitats + altitude ≥ 1800 m a.s.l.' The records refer to 22 Alpine and Pyrenean small lakes, and 3 ponds above 2000 m a.s.l in France (Viets 1913; Migot

1926; Motas 1928a, b; Motas and Soarec 1939; Angelier 1950, 1965; Angelier and Angelier 1953); one small lake at 2000 m a.s.l. in Sierra de Guadarrama, Spain (Valdecasas and Camacho 1986); 7 small Alpine lakes located at 1850–2250 m a.s.l. in Italy (Monti 1903, 1904, 1910) and 10 ponds at 1870–2300 m a.s.l. in Turkey (Özkan 1982, 1988; Özkan and Boyaci 1992). From these habitats, 24 species were reported from mountain lakes and 12 from ponds.

The aim of this paper is to add new data and to enlarge our knowledge on the distribution and ecology of water mites in high elevation mountain lakes to better understand the role played by some ecological factors on the altitudinal and latitudinal distribution patterns in water mites and other freshwater invertebrates. The study is part of an intensive survey of biological and ecological characterization of small Alpine lakes undertaken by the CNR-Institute of Ecosystem Study (Verbania, Italy) in the frame of an UE sponsored project (EMERGE: European Mountain lake Ecosystems; Regionalization, DiaGnostics & Socio-economic Evaluation).

Study area and methods

The lakes investigated (Figure 1) are located at high altitudes (1900–2400 m a.s.l.), above the treeline limit in the Ossola Valley (Central Alps, Piedmont, Italy) and originated from glacial erosion during the last ice age. They are characterized by small size (area lower than 0.2 km^2), low depth (2–15 m), lithological substratum mainly formed by crystalline bedrock (Table 1) and extended ice- and snow-cover (from October to June). Watershed areas are not influenced by direct human impact. Lake selection was based on their distribution along the main geographical and environmental gradients of the Ossola Valley, so that they are characterized by diverse water chemistry (Table 2). The lakes receive acid deposition and evidence of water acidification was found by means of both dynamic modelling (Rogora et al. 2003) and paleolimnological studies (Musazzi and Marchetto 2002).

Sampling was performed during the ice-free season (Table 2). Water mites were qualitatively sampled, together with other benthic invertebrates, by hand-netting and kick-sampling (225 μm mesh size) the littoral zone, inlets and outlets of lakes, when present. In the laboratory the animals were separated from the mud under a stereomicroscope (40×), counted and mounted on slides following standard procedures (Di Sabatino et al. 2000).

Water samples were taken at the same time as biological ones and then analysed in laboratory for major chemical variables (Table 2). For details on sampling procedure and analytical methods followed, see Rogora et al. (2001) and Tartari and Mosello (1997). Water temperature was measured in the field.

Multivariate analysis (principal component analysis (PCA)) was performed on environmental variables and species composition of the water mite community, using the MVSP 3.1 statistical package (Kovach 1995, 1999). In a first step, the lakes were ordered on the basis of their morphometric and

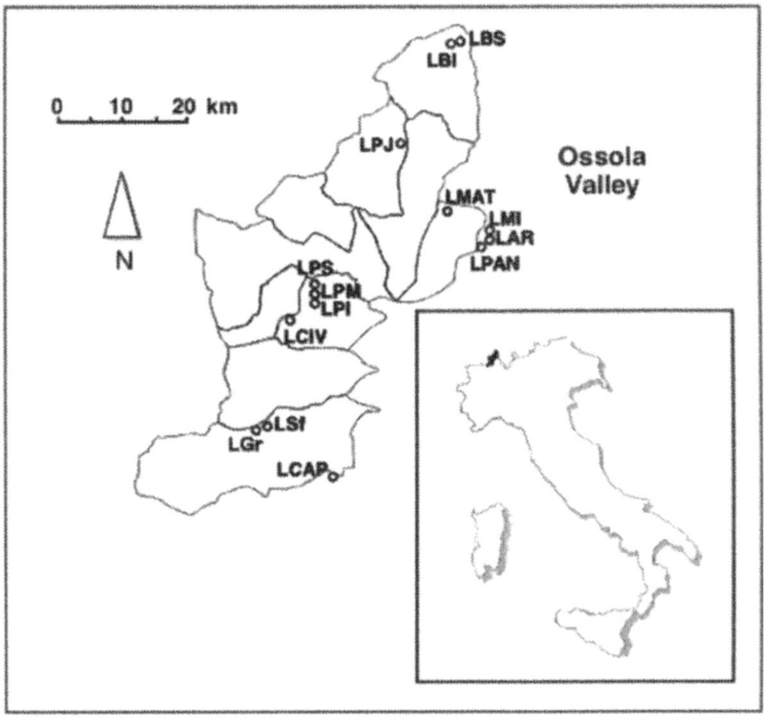

Figure 1. Map of the study area and location of the lakes investigated. Watersheds and lake abbreviations are as in Table 1.

physico-chemical parameters. The following variables were chosen: altitude, lake area, watershed area, depth, temperature, pH, sulphates and total P. Conductivity, calcium content and alkalinity were excluded from the analysis because they are strongly correlated with pH; reactive P and total N were also excluded because of missing data (Table 2). Due to different units of measurements all data were standardized prior to analysis. A second PCA was conducted on lake biological characteristics based on the composition of water mite assemblages (presence–absence); lake LPJ (for lake code see Table 1) had only one species and was excluded from this analysis.

Results

Due to the lithology of the watersheds (mainly formed by crystalline bedrock), most of the lakes investigated are characterized by dilute waters with low conductivity (<25 μS cm^{-1}) and pH ranging from 5.5 to 6.7. The presence of calcareous rocks in the watersheds of the remaining 4 lakes accounts for higher values of conductivity (>40 μS cm^{-1}) and pH (>7.0). Alkalinity is around

Table 1. Geographical and morphometric parameters of the lakes investigated with information on land cover percentages and lithology of the watersheds (presented in decreasing importance).

Lake name:	Alpe Ruggia	Boden inf	Boden sup	Campo IV	Capezzone	Grande	Matogno	Muino inf	Panelatte	Paione inf	Paione med	Paione sup	Pojala	Sfondato
Lake code:	LAR	LBI	LBS	LCIV	LCAP	LGR	LMAT	LMI	LPAN	LPI	LPM	LPS	LPJ	LSF
Valley:	Onsernone	Formazza	Formazza	Bognanco	Strona	Anzasca	Isorno	Onsernone	Onsernone	Bognanco	Bognanco	Bognanco	Devero	Anzasca
Altitude (m a.s.l.)	1890	2334	2343	2279	2100	2269	2087	1910	2063	2002	2147	2269	2305	2422
Lake area (km^2)	0.002	0.052	0.029	0.007	0.009	0.007	0.030	0.002	0.006	0.007	0.007	0.009	0.054	0.005
Watershed area (km^2)	0.182	0.910	0.300	0.710	0.320	0.900	1.360	0.117	0.120	1.260	0.850	0.500	1.070	0.200
Max. depth (m)	2	6.5	6.3	7.4	14	5	4	2	4	13.5	4.9	11.7	15.5	4
Land cover														
Alpine meadows %	28	15	7	2	29	2	65	63	42	8	6	5	25	3
Bare Rocks %	8	80	84	97	69	98	29	25	52	90	92	93	68	97
Shrubs %	61	–	–	–	–	–	4	8	–	–	–	–	–	–
Peat bogs %	–	–	–	–	–	–	–	–	–	–	–	–	2	–
Lithology	Acidic – –	Calcareous Acidic –	Acidic – –	Acidic Calcareous Mafic	Acidic (Calcareous)[a] –	Acidic – –	Acidic Calcareous –	Acidic – –	Acidic – –	Acidic – –	Acidic – –	Acidic – –	Calcareous (Acidic)[a] –	Acidic – –

[a] Rare rock type.

Table 2. Physical and chemical characteristics of the lakes investigated.

Lake code	Sampling date	Temp (°C)	pH	Conductivity (μS cm^{-1} 20 °C)	Calcium (mg l^{-1})	Alkalinity (μeq l^{-1})	Sulphates (mg l^{-1})	Reactive P (μg P l^{-1})	Total P (μg P l^{-1})	Total N (mg N l^{-1})
LAR	18/06/92	11.0	5.81	13	1.15	2	2.65	0	4	ND
LBI	28/09/94	6.0	7.87	52	7.60	468	0.57	ND	5	ND
LBS	28/09/94	5.5	7.75	45	7.40	389	0.81	2	3	ND
LCIV	05/10/00	5.2	7.47	49	7.56	375	5.76	1	2	0.25
LCAP	26/09/00	8.5	6.77	23	3.60	138	2.29	1	4	0.37
LGR	03/10/00	6.4	5.76	7	0.57	0	1.52	1	1	0.25
LMAT	21/09/00	11.7	8.03	80	15.60	761	4.85	1	4	0.26
LMI	18/06/92	12.0	6.04	11	1.09	3	2.19	0	2	ND
LPAN	18/07/91	9.0	6.71	12	1.68	52	1.67	2	5	0.45
LPI	12/09/00	16.0	6.64	13	1.32	38	2.10	1	2	0.35
LPM	12/09/00	13.0	6.53	13	1.18	35	1.99	2	5	0.44
LPS	13/09/00	13.3	6.06	9	0.67	3	1.62	1	6	0.43
LPJ	04/10/00	7.9	7.17	33	4.50	250	3.44	4	6	0.11
LSF	03/10/00	4.2	5.58	9	0.64	0	1.67	1	2	0.34

Lake codes are as reported in Table 1.

zero in 5 lakes, 3 lakes present values below 50 $\mu eq\ l^{-1}$ and 6 lakes have values above 100 $\mu eq\ l^{-1}$ (Table 2). Calcium is the prevalent cation with concentrations of 0.57–7.56 $mg\ l^{-1}$ (only LMAT has values above 15 $mg\ l^{-1}$), while sulphates are among the most prevalent anions (1.52–5.76 $mg\ l^{-1}$) due to atmospheric contribution.

All the lakes investigated are characterized by oligotrophic and ultra-oligotrophic waters. Total inorganic nitrogen and total phosphorus concentrations are below 0.5 $mg\ N\ l^{-1}$ and below 6 $\mu g\ P\ l^{-1}$, respectively. Temperatures vary from 5 to 16 °C; lower values are found in lakes at higher altitude, warmer temperatures are recorded in shallower lakes or in lakes with southern watershed exposure.

Although no lake had pH values lower than 5, some of them (LAR, LGR, LMI, LPS, LSF), with pH < 6 and alkalinity around zero, experience acid pulses during snowmelt and/or heavy rainfall (e.g. Mosello et al. 1999). Three lakes (LPAN, LPI, LPM) with alkalinity values below 50 $\mu eq\ l^{-1}$ can be considered sensitive to acidification; the remaining 6 lakes (LBI, LBS, LCIV, LCAP, LMAT, LPJ) with alkalinity above 100 $\mu eq\ l^{-1}$ are well buffered and do not suffer from acidic deposition.

Seventeen species of Hydrachnidia and one species belonging to freshwater Halacaridae were collected (Table 3). The most abundant species sampled were *Lebertia rufipes*, *Arrenurus conicus* and *Lebertia tuberosa* which were present in the majority of the lakes. *Lebertia sefvei*, *Oxus setosus*, *Feltria minuta*, *Panisus torrenticolus*, *Wettina podagrica* and *Sperchon glandulosus* were also well represented but relatively less abundant. All other species occurred with few individuals and only in single biotopes. *Atractides fissus* and *Arrenurus conicus* are recorded for the first time from Italy.

Among the Hydrachnidia, most of the species sampled show a preference for spring habitats (8 species, 47%); five species (29%) are stenothermic rheobionts well adapted to environmental conditions of first-order streams and only four (23%) are typical of standing waters. *Arrenurus conicus*, *Lebertia porosa*, *O. setosus* and *W. podagrica* were exclusively sampled in the littoral zone of the lakes whereas *F. minuta*, *L. sefvei*, *L. rufipes*, *L. tuberosa*, *P. torrenticolus* and the halacarid mite *Soldanellouyx chappuisi* occurred both in the littoral and in the small outlets and inlets which characterize some of the lakes investigated. The remaining species (*A. fissus*, *A. gibberipalpis*, *A. loricatus*, *Feltria setigera*, *Partnunia steinmanni*, *Protzia distincta*, *Sperchon brevirostris* and *S. glandulosus*) were found only in the outlets and never collected in the littoral zone.

Most lakes are colonized by 3–5 species; only 1–2 species were collected in LPJ, LBI and LSF (Table 3).The highest diversity was recorded in LPM and LPI with 11 and 9 species, respectively. These lakes, together with LPS, lie very close together, are connected by permanent inlets and outlets and form a complex and heterogeneous cascade hydrosystem with a number of different microhabitats.

A PCA was applied to group lakes and environmental variables. The first two PCA axes explained 57.0% of the total variance. Along the first axis, pH

Table 3. Presence and abundance of water mite species in the lakes investigated.

Species	Species code	Lake codes														Total ind
		LAR	LBI	LBS	LC IV	LCAP	LGR	LMAT	LMI	LPAN	LPI	LPM	LPS	LPJ	LSF	
Hydrachnidia																
Arrenurus conicus Piersig, 1894	Acon	3			1	3		1	3	1	2	56	10			80
Atractides fissus Walter, 1927	Afis										3					3
Atractides gibberipalpis Piersig, 1898	Agib			8							2					10
Atractides loricatus Piersig, 1898	Alor										1					1
Feltria minuta Koenike, 1892	Fmin		10	5			1					2				18
Feltria setigera Koenike, 1896	Fse											1				1
Lebertia sefvei Walter, 1911	Lse	8			7			3	1							19
Lebertia porosa Thor, 1900	Lpo									1						1
Lebertia rufipes Koenike, 1902	Lru		13	29		10	3					33		1	3	92
Lebertia tuberosa Thor, 1914	Ltu	4			9	5	1	2		18	1	6			3	49
Oxus setosus Berlese, 1885	Ose	1					3			1		10	2			17
Panisus torrenticolus Piersig, 1898	Pto						2	4	1	2		1				10
Partnunia steinmanni Walter, 1906	Pst											1				1
Protzia distincta Walter, 1922	Pdi										2					2
Sperchon brevirostris Koenike, 1895	Sbr										1	1				2
Sperchon glandulosus Koenike, 1886	Sgl	1				1					1	4				6
Wettina podagrica Koch, 1837	Wpo										1		1			2
Halacaridae																
Soldanellonyx chappuisi Walter, 1917	Sol											2	1			3
Total individuals		17	23	42	17	19	10	10	5	23	14	117	14	1	6	318
Total species		5	2	3	3	4	5	4	3	5	9	11	4	1	2	

Lake codes are as in Table 1.

(alkalinity, calcium and conductivity) seem the most important factors with the lakes LBI, LBS, LCIV, LPJ and LMAT plotted on the right side of Figure 2 (positive scores). These lakes are larger, deeper and well-buffered. Conversely, smaller and shallower lakes (LMI, LAR, LPAN, LGR, and LSF) with low pH and alkalinity values, are plotted on the left (negative scores). Temperature and altitude (inversely correlated) are the most relevant variables for the second axis with the lakes LBI, LBS LGR and LSF showing distinctly negative scores (lower temperature and higher altitude). Positive scores characterize the warmer lakes (LPI, LMAT, LPS and LPM).

A second PCA was applied to the biological data to group the lakes on the basis of the differential distribution of water mite species. The first two PCA axes explained 46.0% of the total variance (Figure 3). Along the first axis the lakes LSF, LGR, LBI, LBS and to a lesser extent LPM have negative scores, all other lakes show positive values. The axis is mainly defined by the vectors of the species *F. minuta* and *L. rufipes* (negative part), and *A. conicus* (positive part). Lake scores are significantly correlated with altitude ($r = -0.74$; $p < 0.005$) and temperature ($r = 0.67$; $p < 0.01$). The axis could reflect a gradient in elevation and temperature or temperature fluctuations. On the second axis LPI and LPM have distinctly positive scores while LCIV LMAT and LMI have negative ones. The vectors of the species *S. brevirostris* and *S. glandulosus* characterize the positive part of the axis. Both species are rare and occur only in the outlets. Axis scores are not significantly related to any

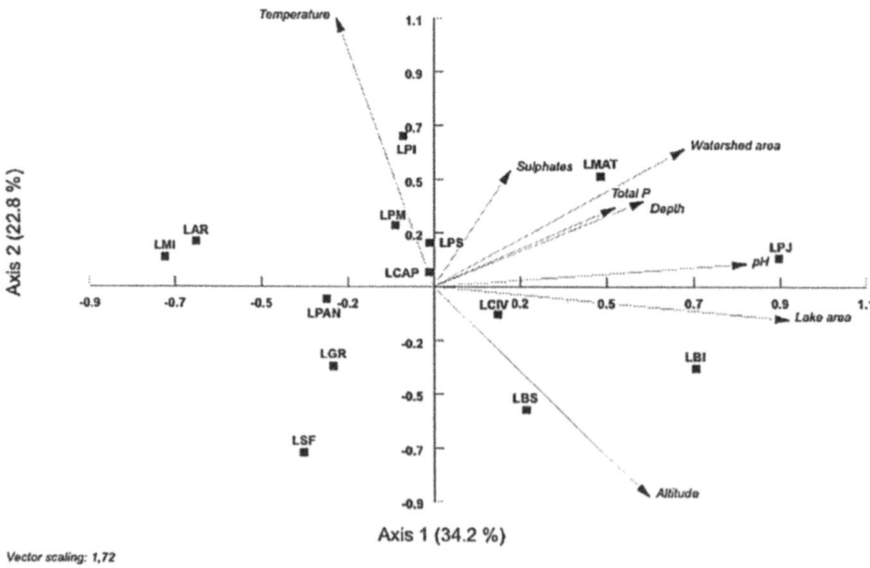

Figure 2. PCA ordination biplot of lake scores and vectors of environmental variables. Morphometric and physico-chemical data were standardized prior analysis. Lake codes are as reported in Table 1.

environmental variable; a positive correlation was found only with lake species richness ($r = 0.84$; $p < 0.001$). Due to the characteristics of LPS, LPM and LPI (see above), the axis could be tentatively interpreted as a gradient of habitat heterogeneity and littoral stability.

Discussion

The lakes investigated are characterized by extreme conditions mainly due to extended ice-cover and ultra-oligotrophy. Among the environmental variables, pH, alkalinity, altitude and temperature are the main factors responsible for their ordination; morphometric parameters are of minor importance. Some lakes are well buffered but others, with low alkalinity and pH, are more vulnerable and could be seriously affected by acidic precipitation.

As reported from running waters (Rundle 1990), water mite assemblages in high Alpine lakes seem little influenced by acidification. Hydrochemistry and morphometry of the lakes are also of minor importance in explaining the differential distribution of water mite species. Conversely, temperature fluctuations, habitat heterogeneity (presence of various microhabitats, presence of permanent outlets and inlets) and the presence of a more stable littoral zone may have a more important role.

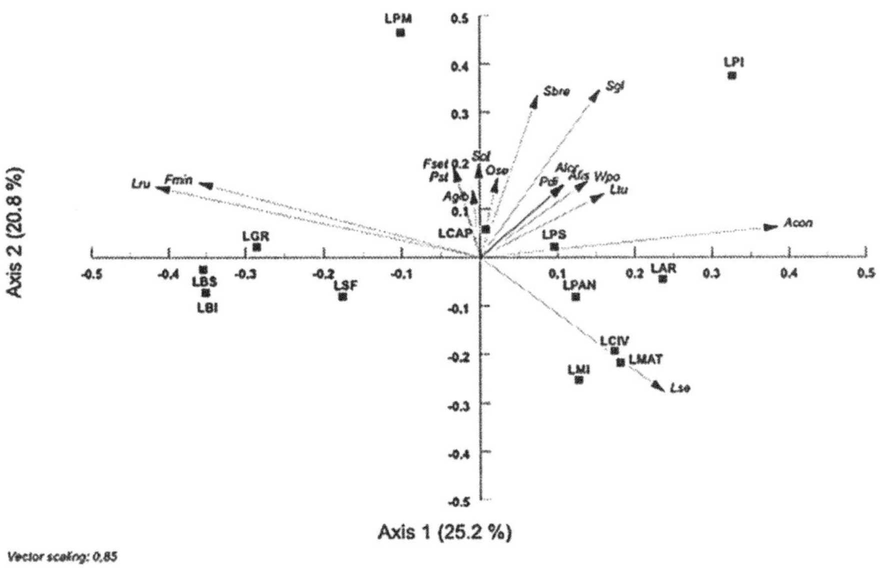

Figure 3. PCA ordination biplot of lake scores and vectors of water mite species (qualitative data). The lake LPJ with only one species was not included in the analysis. Species codes are as reported in Table 3.

Recent research on Alpine European lakes (Fjellheim et al. 2000) have demonstrated the negative impact of acidification on the diversity and distribution of some benthic invertebrates but have also shown that substrate composition and habitat heterogeneity are important factors which allow the presence of more diverse and species-rich invertebrate communities.

Water mite communities in freshwaters are known to be very characteristic and distinctive (Di Sabatino et al. 2000, 2002). A number of species (and genera) is strictly linked to specific habitat-types with only limited shift in habitat preferences (see Gerecke and Di Sabatino 1996). However, in our study, water mite assemblages are mainly composed of stenothermic rheobionts and crenobiontic species. A similar pattern was observed in high altitude lentic biotopes of Central Colorado (Young 1969) and in previous studies conducted on the Alps (Monti 1910; Motas 1928a). In contrast, lakes and ponds at mid to low elevation are characterized by more diverse mite communities (more than 60 coexisting species), composed almost exclusively of limnic taxa (see Pieczynski 1976; Smith et al. 2001; Di Sabatino et al. 2002). The extreme environmental conditions of lentic habitats at high elevations (extended ice cover, low temperatures, low productivity, absence of rooted vegetation) may act as ecological barriers that could limit the presence and the colonization of many eurythermic and eurytopic standing water dwellers. Conversely, for some lotic taxa the littoral zone of high Alpine lakes can simulate the environmental conditions of slow-current sectors of low order streams.

The above observed pattern could also be found in lentic habitats at high latitudes because they are characterized by similar extreme environmental conditions. Lundblad (1962) reported a number of lotic species frequently sampled in lentic habitats in Sweden. However, in Finnish boreal lakes water mite communities, though less diverse in respect to more temperate regions, are mainly composed of standing-water taxa, most of which have their distribution limited to the northern sectors of the palaearctic region (Bagge 1999). Therefore, at a larger spatial scale, other than extant ecological conditions, historical factors may play a determinant role in influencing the diversity and composition of lentic communities.

Other than abiotic factors, some biotic interactions may be important in explaining the altitudinal distribution pattern in water mites. The presence of suitable hosts for parasitic larvae is one of the main constraints in water mites' life cycles. The absence of potential hosts could indirectly explain the absence of limnic water mites in lakes at high elevations. However, in the lakes investigated, Plecoptera, Trichoptera and Chironomidae are well represented but most of the species have their distribution limited to high altitudinal range (Boggero, unpublished).

To conclude, compared to mid-low elevation lentic biotopes, high Alpine lakes are characterized by lower species-richness and by the presence of water mite species typical of springs and first order streams; lentic taxa are rare and represent only a small fraction of the species collected. Extreme environmental conditions are assumed to be the main determinants of the observed pattern. The role of local biotic interactions (presence of suitable hosts for larvae) is less clear and requires further investigation. Altitude, temperature and habitat heterogeneity rather than hydrochemistry and morphometry of the lakes may be more important factors in influencing the distribution and diversity of water mite species but further studies are necessary to confirm our findings.

Acknowledgements

This research was partially funded by the EU project EMERGE (Project No. EVK1-1999-00159) and by the Italian Ministero dell'Ambiente (contract PR4.38/UAQ). Thanks are due to R. Mosello (CNR-ISE, Verbania) for providing chemical data. We also thank H. Proctor (Edmonton, Canada), A. Boulton (Armidale, Australia) and A. Marchetto (CNR-ISE, Verbania, Italy) for helpful comments and stimulating discussion on a first draft of the manuscript.

References

Angelier C. 1950. Hydrachariens nouveaux des Pyrénées. 5eme note. Bull. Mus. Paris. 22: 232–237.

Angelier C. and Angelier E. 1953. Contributions à la connaissance des Hydrachariens des Pyrénées. Le genre *Arrenurus* Dugés. Bull. Mus. Paris. 25: 69–79.

Angelier E. 1965. Les Porohalacaridae de la faune française. Ann. Limnol. 1: 213–220.

Bagge P. 1999. Water mites of small boreal forest lakes (Central Finland). In: Bruin J., van der Geest L.P.S. and Sabelis M.W. (eds), Ecology and Evolution of the Acari. Kluwer Academic Publishers, Dordrecht, pp. 483–489.

Di Sabatino A., Gerecke R. and Martin P. 2000. The biology and ecology of lotic water mites (Hydrachnidia). Freshwat. Biol. 44: 47–62.

Di Sabatino A., Gerecke R., Martin P. and Cicolani B. 2002. Water mites (Hydrachnidia). In: Rundle S.D., Robertson A.L. and Schmid-Araya J.M. (eds), Freshwater Meiofauna: Biology and Ecology. Backhuys Publishers, Leiden, pp. 105–133.

Fjellheim A., Boggero A., Nocentini A.M., Rieradevall M., Raddum G. and Schnell O. 2000. Distribution of benthic invertebrates in relation to environmental factors. A study of European remote Alpine lake ecosystems. Verh. Int. Ver. Limnol. 26: 484–488.

Gerecke R. and Di Sabatino A. 1996. Hystorical zoogeography and evolution of habitat preference in water mites of the Central Mediterranean region. In: Mitchell R., Horn, D.J., Needham G.R. and Welbourn C.W. (eds), Acarology IX, Vol. 1, Proceedings. Ohio Biol. Surv., Columbus, OH, pp. 523–527.

Heino J. 2002. Concordance of species richness patterns among multiple freshwater taxa: a regional perspective. Biodiv. Conserv. 11: 137–147.

Kovach W.L. 1995. Multivariate Data Analysis. Quaternary Res. Assoc., Tech. Guide 5: 1–38.

Kovach W.L. 1999. A Multivariate Statistical Package, Version 3.1 Users' Manual. Kovach Computing Services, London.

Lundblad C.O. 1962. Die Hydracarinen Schwedens II. Ark. Zool. 14: 1–635.

Migot A. 1926. Sur la faune francaise des hydracarides. Bull. Soc. Zool. de France 51: 91–134.

Monti R. 1903. Le condizioni fisico-biologiche dei laghi Ossolani e Valdostani in rapporto alla piscicoltura. Mem. Ist. Lombard., Pavia 20: 51.

Monti R. 1904. Di un'altra nuova specie di *Lebertia* e di alcune idracne nuove per la fauna italiana. Rend. Ist. Lombard. Sci. Lett. 37: 14.

Monti R. 1910. Contributo alla biologia degli idracnidi alpini in relazione all'ambiente. Atti Soc. ital. Sci. Nat. Pavia 49: 167–243.

Mosello R., Marchetto A., Boggero A., Brizzio M.C., Tartari G.A. and Rogora M. 1999. Pluriannual evolution of the hydrochemistry of two Alpine lakes (Lake Paione Inferiore and Lake Paione Superiore, Ossola Valley) in relation to atmospheric loads. J. Limnol. 59: 42–48.

Motas C. 1928a. Contribution à la connaissance des Hydracariens français, particulièrement du Sud-Est de la France. Trav. Labor. Piscicult. Univ. Grenoble 20: 373.

Motas C. 1928b. La faune hydracarienne des eaux courantes et des lacs élevés des Alpes du Dauphiné. C. R. Soc. Biogéographie, Paris 2: 150–186.

Motas C. and Soarec J. 1939. Sur deux nouveaux Hydracariens français recueillis dans les Pyrénées. Ann. Sci. Univ. Jassy 25: 1–13.

Musazzi S. and Marchetto A. 2002. Distribuzione lungo un gradiente di acidità delle diatomee nei sedimenti dei laghi alpini nella Val d'Ossola (Alpi Centrali). Studi Trent. Sci. Nat., Acta Biol. 78: 71–80.

Özkan M. 1982. Wassermilben (Acari, Actinedida) aus der Türkei. Ent. Basiliensia 7: 29–60.

Özkan M. 1988. *Hydryphantes* (s. str.) *crassipalpis* Koenike, 1914 (Hydryphantidae, Hydrachnellae, Acari) üzerine bir arastirma. Doga Tr. J. Zool. 12: 86–110.

Özkan M. and Boyaci Y.Ö. 1992. A new *Forelia* Haller 1882 species (Pionidae, Hydrachnellae, Acari) for the Turkish fauna. Doga Tr. J. Zool. 16: 385–394.

Pieczynski E. 1976. Ecology of water mites (Hydracarina) in lakes. Pol. Ecol. Stud. 2: 5–54.

Rogora M., Marchetto A. and Mosello R. 2003. Modelling the effects of the deposition of acidity and nitrogen on selected lakes and streams in Central Alps (Italy). Hydrol. Earth Syst. Sci. 7: 540–551.

Rogora M., Mosello R., Marchetto A., Boggero A. and Tartari G. 2001. Long-term variations in the hydrochemistry of Alpine lakes in the Ossola and Sesia Valleys (Central Alps) in relation to atmospheric input and climate change. Studi Trentini Sc. Nat., Acta Biol. 78: 59–69.

Rundle S.D. 1990. Micro-arthropod seasonality in streams of varying pH. Freshwat. Biol. 24: 1–21.

Smit H., Gerecke R. and Di Sabatino A. 2000. A catalogue of water mites of the superfamily Arrenuroidea (Acari: Hydrachnidia) from the Mediterranean countries. Arch. Hydrobiol., Suppl. 121: 201–267.

Smith I.M., Cook D.R. and Smith B.P. 2001. Water mites (Hydrachnida) and other arachnids. In: Thorp J.H. and Covich A.P. (eds), Ecology and Classification of North American Freshwater Invertebrates, 2nd ed. Academic Press, San Diego, CA, pp. 551–659.

Tartari G.A. and Mosello R. 1997. Metodologie analitiche e controlli di qualità nel laboratorio chimico dell'Istituto Italiano di Idrobiologia del Consiglio Nazionale delle Ricerche. Documenta Ist. ital. Idrobiol. 60: 160.

Valdecasas A.G. and Camacho A.I. 1986. Las Hidracnelas leniticas de la Sierra de Guadarrama (Acari, Parasitengona, Hydrachnellae). Graellsia 42: 149–160.

Viets K. 1913. Hydracarinen aus Südostfrankreich. Abh. naturw. Ver. Bremen 21: 333–336.

Wetzel R. 2002. Limnology, 3rd ed. Academic Press, San Diego, CA.

Young V. 1969. Ecological distribution of Hydracarina in North Central Colorado. Am. Mid. Nat. 82: 367–401.

GPSR Compliance

The European Union's (EU) General Product Safety Regulation (GPSR) is a set of rules that requires consumer products to be safe and our obligations to ensure this.

If you have any concerns about our products, you can contact us on ProductSafety@springernature.com

In case Publisher is established outside the EU, the EU authorized representative is:

Springer Nature Customer Service Center GmbH
Europaplatz 3
69115 Heidelberg, Germany

Batch number: 10584688

Printed by Printforce, the Netherlands